T0230453

SpringerBriefs in History of Science and Technology

The *SpringerBriefs in the History of Science and Technology* series addresses, in the broadest sense, the history of man's empirical and theoretical understanding of Nature and Technology, and the processes and people involved in acquiring this understanding. The series provides a forum for shorter works that escape the traditional book model. SpringerBriefs are typically between 50 and 125 pages in length (max. ca. 50.000 words); between the limit of a journal review article and a conventional book.

Authored by science and technology historians and scientists across physics, chemistry, biology, medicine, mathematics, astronomy, technology and related disciplines, the volumes will comprise:

1. Accounts of the development of scientific ideas at any pertinent stage in history: from the earliest observations of Babylonian Astronomers, through the abstract and practical advances of Classical Antiquity, the scientific revolution of the Age of Reason, to the fast-moving progress seen in modern R&D;
2. Biographies, full or partial, of key thinkers and science and technology pioneers;
3. Historical documents such as letters, manuscripts, or reports, together with annotation and analysis;
4. Works addressing social aspects of science and technology history (the role of institutes and societies, the interaction of science and politics, historical and political epistemology);
5. Works in the emerging field of computational history.

The series is aimed at a wide audience of academic scientists and historians, but many of the volumes will also appeal to general readers interested in the evolution of scientific ideas, in the relation between science and technology, and in the role technology shaped our world.

All proposals will be considered.

Alessandro De Angelis

Galileo and the 1604 Supernova

With a Translation of the "Dialogo de Cecco di Ronchitti da Bruzene"

Alessandro De Angelis 🆔
Dipartimento di Fisica e Astronomia
University of Padua
Padua, Italy

ISSN 2211-4564   ISSN 2211-4572 (electronic)
SpringerBriefs in History of Science and Technology
ISBN 978-3-031-59485-4   ISBN 978-3-031-59486-1 (eBook)
https://doi.org/10.1007/978-3-031-59486-1

Translation from the Italian language edition: "Galileo e la supernova del 1604" by Alessandro De Angelis,
© Castelvecchi 2022. Published by Castelvecchi. All Rights Reserved.

This Springer imprint is published by the registered company Springer Nature Switzerland AG
The registered company address is: Gewerbestrasse 11, 6330 Cham, Switzerland

Preface

Only seven supernovae seen with the naked eye in the Milky Way are documented; the last one, in 1604, changed the history of astronomy and cosmology. The appearance of a new star—today we know that it is rather a dying star—surprised astronomers and astrologers, who believed that stars were fixed and unchangeable. Scientists with different conceptions of the Universe competed and collaborated to explain its nature and origin, studying its astrological signals; among them Galilei and Kepler, but also Arab, Chinese, and Korean astronomers. In 1604, observations could only be made with the naked eye: the telescope would be invented only 5 years later. What remains of the explosion is still visible today, and the data recorded by our large ground-based and satellite-based observatories complement the measurements of astronomers of the beginning of seventeenth century.

In 1604 Galileo was teaching the mechanics of the planets in Padua, and was asked his interpretation of the event. He was very careful in expressing his opinions. He did it with three lectures that he never published, with a pseudonymous treatise in Paduan dialect, and with a poem immediately withdrawn. A quarrel among Galileo and prominent Aristotelian scholars working in Italy at those times started, which dominated the scientific debate in 1604–1606 (before the appearance of Kepler's treatise *De stella nova*.) In this book we reproduce the most important documents of the discussions in Italy.

Since 1604 there has never been a galactic supernova visible to the naked eye. Nobody can say when and where the next one will appear. Perhaps it will be Betelgeuse, the brightest star in the constellation of Orion: this giant red star not far from us will explode at an unspecified time between now and one hundred thousand years in the future, and will become one thousand times brighter than Venus. We physicists would like this to happen tonight.

Padua, Italy   Alessandro De Angelis

A Tranquil Star

Once upon a time, somewhere in the Universe very far from here, lived a peaceful star, which moved peacefully in the immensity of the sky, surrounded by a crowd of peaceful planets about which we have not a thing to report. This star was very big and very hot, and its weight was enormous [...].

This tranquil star wasn't supposed to be so tranquil. Maybe it was too big: in the far-off original act in which everything was created, it had received an inheritance too demanding. Or maybe it contained in its heart an imbalance or an infection, as happens to some of us [...].

Of this restlessness Arab and Chinese astronomers were aware. The Europeans, no: the Europeans of that time, which was a time of struggle, were so convinced that the heaven of the stars was immutable, was in fact the paradigm and kingdom of immutability, that they considered it pointless and blasphemous to notice changes. There could be none—by definition there were none. But a diligent Arab observer, equipped only with good eyes, patience, humility, and the love of knowing the works of his God, had realized that this star, to which he was very attached, was not immutable. He had watched the star for 30 years, and had noticed that it oscillated between the fourth and the sixth of the six magnitudes that had been described many centuries earlier by a Greek, who was as diligent as he, and who, like him, thought that observing the stars was a route that would take one far. The Arab felt a little as if it were his star: he wanted to place his mark on it, and in his notes he called it al-Ludra, which in his dialect means "the capricious one." [...]

An observer who, to his misfortune, found himself on October 19th of that year, at 10 o'clock our time, on one of the silent planets of al-Ludra would have seen, "before his very eyes," as they say, his gentle sun swell, not a little but "a lot," and would not have been present at the spectacle for long. Within a quarter of an hour he would have been forced to seek useless shelter against the intolerable heat—and this we can affirm independently of any hypothesis concerning the size and shape of this observer, provided that he was constructed, like us, of molecules and atoms—and in half an hour his testimony, and that of all his fellow-beings, would end. Therefore, to conclude this account we must base it on other testimony, that of our earthly instruments, for which the event, in its intrinsic horror, happened in a "very" diluted form and, besides, was slowed down by the long journey through the realm of light that brought us the news. After an hour, the seas and ice (if there were any) of the no longer silent planet boiled up; after three, its rocks melted and its mountains crumbled into valleys in the form of lava. After ten hours, the entire planet was reduced to vapor, along with all the delicate and subtle works that the combined labor of chance and necessity, through innumerable trials and errors, had perhaps created there, and along with all the poets and wise men who had perhaps examined that sky, and had wondered what was the value of so many little lights, and had found no answer. That was the answer.

After one of our days, the surface of the star had reached the orbit of its most distant planets, invading their sky and, together with the remains of its tranquillity, spreading in all directions—a billowing wave of energy bearing the modulated news of the catastrophe.

> From Primo Levi, *Una stella tranquilla*
> La Stampa, 29 gennaio 1978
> Translation from *The New Yorker*, February 4, 2007

Contents

Chapter 1
Supernovae

On the evening of October 9, 1604, near that place in the sky of a close conjunction between Mars, Jupiter, and Saturn, a new light, brighter than all the planets with the exception of Venus, suddenly appeared. It remained there for a year and a half and then it disappeared after slowly fading. That *stella nova* (new star), as it was called, changed the history of astronomy and cosmology: it was believed at that time that stars were fixed, immutable and ungenerable, but obviously this was not true. Scientists with different conceptions of the Universe competed and collaborated to explain its nature and origin; among them Galilei and Kepler, the two most famous "astrophysicists" of the time, but also Arab, Chinese and Korean astronomers. Today we know that the new star was a supernova, the last of the seven supernovae observed with the naked eye in the Milky Way—the six previous ones had been recorded in the years 185, 393, 1006, 1054, 1181, 1572—of which documentation exists. So not a new star, but a star that dies and explodes.

Stars and binary star systems formed by two stars orbiting each other (which constitute about one third of the "stars" we observe), if they are "heavy" enough, end their life by gravitationally collapsing on themselves and releasing a large amount of energy. These collapse phenomena are called "supernovae", a term coined by the German astronomer Walter Baade and the Swiss Fritz Zwicky in 1934—the term "nova", instead, denotes in modern astronomy a great increase of the brightness of a star that then returns to its original state in a time that typically goes from some weeks to some months; this phenomenon, due in the "classical" case to the accretion of a star at the expense of its partner in a binary system, has a typical energy release much smaller than that of a supernova, and can be recurrent—the star continues to exist and can give rise to other explosions as soon as it can be replenished by material taken from the secondary star. The term *stella nova* was invented by the Danish Tycho Brahe (astronomer of the Holy Roman Emperor, predecessor and teacher of Kepler) to describe the astronomical event of 1572 that today we know is actually a supernova—we call it "Tycho's supernova".To make a long story short, astronomers

© The Author(s), under exclusive license to Springer Nature Switzerland AG 2024
A. De Angelis, *Galileo and the 1604 Supernova*,
SpringerBriefs in History of Science and Technology,
https://doi.org/10.1007/978-3-031-59486-1_1

before the twentieth century did not distinguish between the two phenomena. Brahe however also studied what some think is the first nova of which there is a track in the scientific literature, appeared in 1600 in the constellation of Cygnus—others think it is a different type of star with variable luminosity, with a different mechanism from the "classical nova" that I described before.

A supernova close enough to Earth is a spectacular event: it can appear as a new astronomical object as bright as Venus, or even more, visible for several months. What does close enough mean? The Earth and the solar system and almost all the stars we see with the naked eye belong to a single galaxy, the Milky Way, the galaxy par excellence; this is one of the galaxies of the Universe, which are estimated to be about one hundred billion. The name of the Milky Way (from Latin Via Lactea) derives from Greek galaxias, etymologically related to the word milk, used in Greek times to designate it—the term refers to its appearance as a faint whitish band of light with a milky appearance that crosses diagonally the sky. The Milky Way is brightest in the direction of the constellation Sagittarius, where the galactic center is located, which is about one hundred thousand light years away from us.[1] The nearest stars (such as those of the constellations that dominate the sky: Orion, Cassiopeia, ...) appear distinct from the Milky Way for a pure geometric effect related to their proximity; this was not understood by the ancients, who thought they were not in the Galaxy. Only since 1930 we have understood that our galaxy is not the only one, even if Kepler had already fomented this hypothesis. When we talk about a close supernova we mean that its distance is about twenty thousand light years, or less.

Supernovae are classified from the point of view of their appearance into two "types". If the spectrum of emitted light contains emission lines typical of hydrogen (the initial fuel of a star) the supernova is called of type II; otherwise it is called of type I. In each of these two types there are subdivisions depending on the presence of signatures of other elements or on the shape of the light curve (the graph of the apparent luminosity of the supernova as a function of time): Ia, Ib, etc.

The brightness of an astronomical object seen by an observer on the Earth is called apparent magnitude (or also "magnitude"): a brighter object appears larger to us. The magnitude scale has its origin in the Hellenistic practice of dividing stars into six magnitudes: the brightest stars were called of first magnitude (magnitude 1), while the faintest were of sixth magnitude (magnitude 6), the limit of human visibility with the naked eye. The system is now formalized by defining a star of first magnitude as a star that is 100 times brighter than a star of sixth magnitude. The scale is "harmonic," that is, the relationship between two consecutive brightnesses is constant: therefore, a star of first magnitude is about 2.5 times brighter than a star of second magnitude (obviously the brighter an object appears, the lower the value of its magnitude). The magnitude 0 is approximately that of the stars Arcturus and Vega. The planet Venus has an apparent magnitude of -4.2.

---

[1] A light year is the distance covered by light in one year, about 10^{16} m (ten thousand billion kilometers), which is 63 thousand times the distance between the Earth and the Sun. One parsec, a unit of distance used by astrophysicists, corresponds to about 3.3 light years.

In addition to their observed properties, supernovae can also be classified by the dynamics of the explosion.

1. Supernovae resulting from core collapse (types II, Ib, Ic). At the beginning, stars burn hydrogen in their core, producing heavier elements. If the star is large enough, when the hydrogen is exhausted, the core contracts under the weight of the star itself until the density and temperature reach such conditions that the heavier elements start to fuse generating nuclear energy. This sequence (fuel exhaustion, contraction, heating and ignition of the ashes of the previous cycle) can repeat itself several times depending on the mass, finally leading to a explosive combustion. Most of the gravitational energy of about 10^{46} joule released in the collapse (an enormous energy, equal to one million billion billion times the energy consumed by we all Terrestrials in one year) is released, mainly in the form of invisible particles called neutrinos, in a burst that lasts a few seconds and whose debris expand for millennia at decreasing speed.

2. Type Ia (read "type one a") supernovae occur when, in a binary system formed by a white dwarf (a dense, inactive star of about the same mass as the Sun) and another star, the white dwarf "vampirizes" its companion reaching a total critical mass of about 1.4 solar masses. Beyond this mass, the system may collapse and trigger an uncontrolled thermonuclear explosion. Type Ia supernovae seem to have approximately the same dynamics and the same energy emission. The supernova of 1604 is believed on the basis of its behavior to be a type Ia supernova (Fig. 1.1).

A *supernova remnant* is the structure formed after the explosion of a supernova: a compact object, such as a neutron star (which has a density of one billion tons per cubic centimeter), or even a black hole, can form at the center of the exploded star, while the material ejected after the explosion appears as a bubble of hot expanding gas that shakes and sweeps the surrounding interstellar medium. This process can accelerate the particles trapped in the structure (mainly hydrogen nuclei) bringing them to very high energies: it is believed that supernovae are at the origin of most of the cosmic rays (high-energy particles arriving from the extraterrestrial regions of the Universe to the Earth) in the Galaxy.

Together with Kepler, Galileo Galilei (for his writings I refer where possible to the national edition [1], briefly called *Opere* in the following, edited by Antonio Favaro) gave the greatest contributions to the study of the supernova of 1604; at that time the two, who corresponded by exchanging letters even if apparently they did not do so at crucial moments of the supernova explosion, were the best known astronomers in the world.

Although Galilei's life is well known, it is useful to make a brief summary under-lining some aspects useful to understand the role of this scientist. Galileo Galilei (1564–1642), who taught in Padua from 1592 to 1610 (and therefore also during the observation of the supernova), gave fundamental contributions to physical sci-ences, astronomy and to the development of the scientific method. His discoveries with the telescope revolutionized astronomy and paved the way for the acceptance of the Copernican heliocentric system; his defense of this system and his attack on the

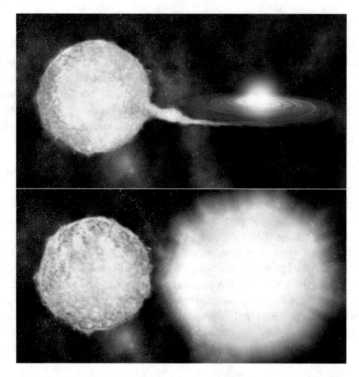

Fig. 1.1 Schematic description of a type Ia supernova. Top: the white dwarf is accreting. Bottom: having reached its critical mass, it explodes. Credits: NASA/Chandra

Ptolemaic system and of the dogmatic view of nature embodied in Aristotelianism eventually led to his notorious condemnation by the Inquisition.

The *stella nova* was a challenge to Aristotle's physics and cosmology, which were dominant at the time. Aristotle (384–322 BC), from Stagira in Northern Greece, formerly a disciple of Plato, dominated the scientific thought from the late classical greek period, covering with his studies subjects including physics, astronomy, biology, zoology, metaphysics, logic, ethics, aesthetics, poetry, theatre, music, rhetoric, psychology, linguistics, economics, politics, meteorology, geology, and politics. Aristotle provided a synthesis of the various philosophies existing prior to him. He designed the intellectual lexicon of western civilization, and his philosophy has exerted a unique influence on almost every form of knowledge—in particular in the scientific method before the Galilean revolution.

In Aristotle's cosmological system, the stars and planets rotate on invisible spheres, located one inside the other and centered on the Earth. The elements (air, earth, water, fire), who are subject to changes, are located between the Moon and the Earth; the region beyond the Moon is made by an incorruptible substance or "fifth essence", not subject to the cycle of generation and corruption: thus there can be no change beyond the sphere of the Moon. Ptolemy[2] had perfected the description by adding more complex mathematical concepts and secondary motions, which were able to justify the retrograde motions of the planets—sometimes the planets reverse their direction of motion with respect to the Earth—and their approximately uniform rotation around the Sun. And to remedy the fact, also discovered already by Hipparchus,[3] that even the so-called fixed stars seem to have a slow irregular motion (and also for a fashionable religious interpretation), after Ptolemy a ninth sphere (in addition to those of the Moon, the Sun, the planets Mercury, Venus, Mars, Jupiter, Saturn, and the eighth, that of the fixed stars or "firmament") was introduced, external to everything, called "primum mobile".

In Aristotle's doctrine the Sky and the Earth were thus absolutely distinct, and any kind of change could occur only in the region below the Moon, that is on the Earth or in the circumterrestrial medium. The supernova of 1604 appeared from nowhere, exceeding in brightness all stars and planets except Venus. This should have been impossible for a star, and some scientists preferred to see it as a meteorological event in the upper atmosphere. But the measurements established that the new object was beyond the sphere of the Moon, therefore part of the Sky considered immutable.

Supernovae are identified by a code combining the prefix SN (which stands for supernova), the year of discovery, and, if necessary, a one- or two-letter suffix (but until the advent of modern astronomy no more than one supernova was ever detected per year).

What remains of the supernova of 1604, today called SN 1604 or "Kepler's supernova", is still visible with appropriate astronomical instruments. The German astronomer Walter Baade (1893–1960) in 1943 associated a nebula at a distance of about 5 thousand parsecs, that is 16 thousand light years, to that astrophysical relict. With the progress of infrared astronomy and X-ray astronomy, occurred in the second half of the twentieth century, the nebula could be observed in detail and we can study its morphology. It has become apparent that the rest of the supernova is still expanding, at a speed of about 10,000 km per second on average (very large for a supernova of this age). SN 1604 has become again a battlefield for astrophysical theories: this type of supernova can be a site for the acceleration of cosmic rays (particles of extraterrestrial origin with very high energies, even billions of times larger

---

[2] Claudius Ptolemy (c.?100—c.?170 AD) was a mathematician, astronomer, astrologer, geographer, and music theorist. He lived in or around the city of Alexandria, in the Roman province of Egypt under Roman rule; several historians think that he was also a Roman citizen. He is famous in particular for his astronomical treatise now known as the Almagest (The Greatest Treatise). Ptolemy presented his astronomical models alongside convenient tables, which could be used to compute the future or past position of the planets. The Almagest also contains a star catalogue, which is an updated version of the catalogue created by Hipparchus.

[3] Hipparchus of Nicaea (c.?190—c.?120 BC) was a Greek astronomer, geographer, and mathematician, and is considered the greatest astronomical observer of antiquity. His reputed achievements include the discovery and measurement of Earth's precession, and the compilation of the first comprehensive star catalog of the western world.

than those produced by terrestrial accelerators such as the one at CERN in Geneva), and its emission is studied using telescopes sensitive to different wavelengths, up to the extreme energies of gamma rays.

Reference

1. *Le Opere di Galileo Galilei,* Edizione Nazionale, edited by Antonio Favaro, vol. I–XX, Barbera, Firenze 1890–1909

Chapter 2
The 1604 Supernova

On the evening of October 9, 1604, a conjunction between Mars, Jupiter, and Saturn was expected as they passed in front of the constellation of Sagittarius.

Many eyes of amateur and professional astronomers all over the world were pointed to the sky (let us remember that the astronomical telescope will be invented only five years later), because astrologers had predicted great events following this rare conjunction. The conjunction between Jupiter and Saturn occurs approximately every twenty years; every 200 years three consecutive conjunctions also bring Mars closer, and this was the case in 1604. In astrology the ecliptic, i.e., the circumference along which the Sun apparently moves, is divided into twelve "zodiacal" signs (from the Greek zoidiakos, derived from zoidon, "figure of an animal"), associated three by three to the four material elements of Greek physics: earth, water, air and fire. By joining three signs with an equilateral triangle, you have four "trigons" that are each associated with an element. The supernova of 1604 appeared in the "fire trigon" which contains Aries, Leo and Sagittarius, that is the first, the fifth and the ninth sign. A great conjunction in a fire trigon occurs once every 800 years.

In correspondence with such an event astrologers predicted important consequences for humankind—wonderful events had already happened in the past. Here is a collection of such events made by Kepler in his *De stella nova*.

Year	Since the beginning of the world	Key people	Associated events
4000 a.C	0	Adam	Creation of the world
3200 a.C	800	Enoch	Robberies, cities, arts, tyranny
2400 a.C	1600	Noah	Flood
1600 a.C	2400	Moses	Exodus from Egypt. Law
800 a.C	3200	Isaiah	Era of the Greeks, Babylonians, Romans
0	4000	Jesus Christ	Roman dominance, Renewal of the world
800 d.C	4800	Charlemagne	Western Empire And Saracen Empire

© The Author(s), under exclusive license to Springer Nature Switzerland AG 2024
A. De Angelis, *Galileo and the 1604 Supernova*,
SpringerBriefs in History of Science and Technology,
https://doi.org/10.1007/978-3-031-59486-1_2

An unexpected guest participated to the celestial event, close to the joined planets: a new star.

The first observations of the supernova in Italy were reported the same day in Verona by Ilario Altobelli (1560—1637), a friar from Macerata, astronomer and from 1599 to 1605 rector and professor of mathematics in the *Studio* of Verona, in Florence by Raffaello Gualterotti,[1] and in Paterno Calabro near Cosenza by an anonymous doctor (who wrote about the observation to Christopher Clavius[2]).

Altobelli described the new star as similar to the stars of the eighth sphere, that is, of a non changing and sparkling color. He undoubtedly placed it among the fixed stars, and described its color as "half yellow, half orange green, or even a mixture of yellow and green". The day when Altobelli saw the star for the first time, it seemed to him that it had the same size of Jupiter, but then, due to bad weather, he could not see anything more until October 15, when the sky finally became clear and the new star reappeared, as he reported, a little bigger than he remembered. In the following days he seemed to see it decrease very slightly in size, but still comparable to Jupiter.

Altobelli expressed a first hypothesis about the composition of the star, starting from the fact that having appeared in a fire sign, close to the ecliptic and not far from the Sun, it could have a nature of burning flame. He also excluded that it could be a sort of globe suspended from a hypothetical equilibrium between layers of air of different density, since the air was considered liquid and mobile, while the star did not show any movement.

The evening when the supernova appeared, as often happens, in Padua the sky was cloudy, and it was not possible to make any observation. The first observation in Padua was made a day later by the students Baldassarre Capra (from Milan) and Camillo Sasso (from Calabria), and by the German teacher Simon Mayr[3] (who had Latinized his name to Simon Marius), from a point of observation presumably corresponding to what today is called by the Paduans "basso Isonzo," about 2 km south-west of the Basilica of St. Anthony. Also the Florentine Lodovico Delle Colombe[4] observed the astronomical event in the early days, but reported the observation only much later.

Due to bad weather conditions, it was only on the 15th and in the following days that Capra and his friends were able to see the star again, bigger than before, and

---

[1] Raffaello Gualterotti (Florence, 1544–1638) was an Italian poet and astronomer. He wrote the *Celebrations in the wedding of Francesco de' Medici Grand Duke of Tuscany and his consort Mrs. Bianca Cappello.*

[2] Christopher Clavius (1538–1612) was a German Jesuit, mathematician and astronomer, then rector of the Collegio Romano, the Jesuit university in Rome.

[3] Simon Mayr (1573—1625) was a German astronomer. He was tutor in Padua from 1601 to 1605, and had bad relations with Galilei, who called him a "poisonous reptile". Later he had a dispute with Galilei on the priority of the observation of Jupiter's moons.

[4] Lodovico Delle Colombe (1565—probably 1623) was a Florentine philosopher and poet; he opposed Galilei in matters relating to astronomy, physics, and cosmology. A year after the works of Capra and Galilei, in 1606, published a book in which he defended an Aristotelian view of cosmology after Galilei had used the occasion of the supernova to challenge the Aristotelian system. Delle Colombe claimed to have observed the star nova on October 9 or shortly after.

larger than all planets except Venus. They carefully noted its color because this could indicate its astrological powers.

Galilei heard about the supernova only after a few days. The news was communicated to him by his neighbor Alvise Cornaro (Galileo and Cornaro's gardens were attached one to the other), who in turn had received it from Capra. Partly because of bad weather conditions, the first observation of Galilei is dated October 28.

In October 1604 Galilei was teaching Mathematics and Astronomy at the University of Padua, and for the academic year that was ending he had chosen the motion of the planets as the topic of his lectures [1–3]. When the new star appeared in the sky, it was therefore the reference figure to which to address for the doubts and questions that such an apparition brought with it.

The new star, bright and pulsating, had generated wonder, but also terror and curiosity. Galilei was officially asked to outline the situation by exposing in some public lectures his point of view, so as to answer the many questions that raged among the ranks of the academic community and among ordinary people. Between November and December 1604 Galilei devoted three public lectures to the new star, which were attended, according to his memoirs, by more than a thousand students and citizens (a fact hard to believe, since the largest classroom in Padua at that time, the Aula Magna, now called Aula Magna Galileo Galilei, has about 400 seats). The lectures were followed by lively debates, variously described as "pleasant discussions" or as "bitter disputes" by contemporary authors. Of these lectures we have the notes relating to the *incipit* (in Latin, translated here):

> On the tenth of October of this 1604, a new light was seen for the first time in the sky; initially weak, but later, after a few days, greatly increased to exceed all the stars, both fixed and mobile, with the exception of Venus alone; a very beautiful and sparkling light, to the point of seeming in the vibrating brightness almost to go out and immediately turn back on; light that exceeds in splendor that of all the fixed stars, including Sirius itself; similar, for the color of light, to the golden splendor of Jupiter and to the reddish color of Mars.

> While in fact it contracts its fearsome rays and gives the wrong idea of a quenching, it appears almost incandescent due to a red Mars-like color, but while it spreads its rays wider, as if it lived again, it shows itself shining with the glow of Jupiter. Someone would be led to believe that this light was generated by the conjunction of Jupiter and Mars; and this, moreover, above all, because it appears to be generated almost in the same place and at the same time in the conjunction of these planets.

> On October 9, in fact, at 5 in the morning, the conjunction of Jupiter and Mars took place in front of Saturn, located only 8 degrees from them, towards west; at which time, observing this conjunction, we saw no other star in that area beyond the three mentioned. The following evening, that is, October 10 at sunset, we saw first this new light; while initially appearing weak and small, soon in a few days it reached a considerable size. It would not be out of place to suppose that this new light was generated at the time of the aforementioned conjunction and, given its tenuity, remained hidden.

> And also when the planets found themselves together at the 19th degree of Sagittarius, in the 18th grade of the same sign this light appeared again; [and a conjunction] was constituted by these four lights.

> This splendor caused even the eyes of the most obtuse people to be raised to the divine realities, as if a new miracle from heaven had happened. What the conjunction of the most splendid and innumerable stars of which the fields of the sky are adorned does not succeed

in carrying out: the condition of human life is in fact such that the daily realities, even those worthy of admiration, escape us; on the contrary, if something unusual happens and outside the norm, this attracts everybody.

You are witnesses, young people who have flocked here to hear me talk about this apparition worthy of admiration; some, frightened and shaken by inconsistent superstition, to understand if the prodigious prodigy announces a bad omen; others wondering if there is a true star in the skies or a boiling steam in the vicinity of the earth; all, then, anxiously trying to know the substance, the motion, the place and the reason for that apparition with unanimous interest. Wonderful desire, gosh, and worthy of your minds!

Heaven willing that the paucity of my intelligence may respond to the importance of the matter and to your expectations. I do not hope or mistrust: I think that I am going to establish only those things that are strictly within my competence, and to report things demonstrated about the movement of matter, so that you will all learn to know them.

and some notes, partly in Latin and partly in the vernacular (here translated from *Opere,* vol. 2). Favaro attempts a chronological order.

On October 28, 1604, I observed the star, which was on the line joining the tail of the Big Dipper and the constellation of the Corona Borealis.

The brightness of a body tends to disappear much faster than the sight of it: during the day we can see a flashlight placed at a great distance even if its light does not illuminate the bodies around us.

The new star did not originate from an explosion, because objects that catch fire very quickly, tend to extinguish and consume just as quickly.

Now it remains to find out finally what I think of this wonderful apparition; I am aware that I have to submit to censorship, and there will be those who will want to accuse me.

The gases that are around the Earth tend to rise upwards, and once far away, reflect the light of the Sun; this light is then released during the night, and returns to the earth, generating an intense twilight. This is a phenomenon I have often observed, and such twilight always appears towards North.

The reason is obvious: at noon, regardless of whether they rise from east or west, such gases are always within the cone of shadow, and then rising to the north end up including the area where we are, and can therefore be observed [...].

I saw Venice around 2 a.m., and the air to the north was very clear, and there were like internal walls of light, almost brighter than the light of the Moon itself. The streets heading north were completely illuminated by that very strong twilight, exactly as if there had been simultaneously the light of the Sun and the Moon [...]. Often such gases appear blood red and yellowish.

Some flames are so weak that, although appearing bright at a distance, they are practically invisible up close: in the same way, the new Star could be some kind of luminous exhalation, which would be invisible up close, while it would be seen from a distance at night in the form of luminous vapors.

The condensation of the ether is so thin that it allows to reflect perfectly the very faint light of the Sun, just as do for example the clouds that cover the tops of the mountains, and that are illuminated by the Sun, the Moon and all the stars. In the same way, the condensed ether is so thin that it would be illuminated even by the smallest of stars.

Regardless of what is the source of light in the heavens, the first big mistake is to consider that all objects react in the same way. In fact, the way the light, for example, of the Sun is absorbed and reflected changes according to the material: for example wood, stone, clouds, or even gold, lead, wood, marble, ice, a fruit, meat, bones, wax, are all materials that react differently. What they have in common is the difference from the clouds, which, once illuminated by the Sun, tend to rise upwards.

This is a new star that shines like the others, but this does not mean that it must be necessarily made of solid matter [...]: in fact, it could be composed of a solid or light body, like the body of the clouds.

There can be a sort of evaporation from the Earth, so big that its quantity is such to constitute a new star [...]: in fact we see every day how the air rapidly fills up with clouds. For example, from a very young wood that is set on fire, the resulting fire is insignificant, while a large amount of smoke is produced.

During his conferences Galilei let it be understood that he had personally made some of the observations actually made by Capra. This fact will cause an unpleasant quarrel with Capra himself, who felt deceived.

In his lectures Galilei drew a very important conclusion, based also on the observations of Spanish and Neapolitan astronomers whose names he did not mention (and later on the observations of Kepler and other Northern European astronomers) with the so-called parallax technique: the new star was beyond the Moon. Parallax is the apparent displacement of an object due to a change in the observer's point of view. If an observer looks at the tip of his nose with the left eye closed, then with the right eye closed, his nose will appear to move against the background; knowing the distance between the eyes we can determine the distance of the tip of the nose from the line joining the eyes.

It is possible to measure the distance of an astronomical object by looking at it from places as far away as Padova and Prague (this was the case for Galilei and Kepler), and to measure its position with respect to the fixed stars from two different points on the Earth. With the instruments of Galileo's time, the stars appeared distant, without parallax: all the observatories measured the same position. If we believe that the Earth orbits around the Sun (and this was the case for Galilei and Kepler), a nearby star will seem to move with respect to the stars more distant if one measures its position six months later, when the Earth lies at opposite points in its orbit around the Sun. The appearance of a new body outside the Earth-Moon system challenged the traditional belief, embodied in Aristotle's cosmology, that the matter of the planets beyond the Moon was unalterable and that nothing new could occur in the Heavens, i.e., beyond the Moon. Galileo was later asked to publish the contents of his lectures. However, he preferred not to do so: "I realize", he wrote, "how weak my arguments are". From the last week of November until after Natale, the new star was too close to the Sun to be visible.

In the meantime a correspondence began between Galilei and other astronomers on observation (the numbering of the letters refers to that in the *Opere*). Having learned of the scheduling of Galilei's lectures, Altobelli wrote him a letter.

(106) Ilario Altobelli to Galileo Galilei
Verona, 3 November 1604

[...] I am happy that You have noticed this new monster in the sky that is driving the Peripatetics crazy. If with the star of 1572 they could afford to assert many untrue things, in front of this star they cannot deny the evidence. Without bothering the parallax, the star is located south, near the ecliptic, in a sign of fire, near Jupiter and not far from the Sun.

It appeared, more beautiful than ever, on October 9 (and not before) with Jupiter and Mars in conjunction. On October 8, in fact, I was observing with a colleague that part of the sky, trying to check if the conjunction between Jupiter and Mars was really respecting the expected positions, and I could see well, thanks to the clear sky, that no star was visible, neither near nor far. Only the three planets were visible. Since many people are asking me about it, I am writing a short description about it, which I will probably finish in eight days. I ask Your Excellency, if you see the star changing its aspect and size, to please let me know, as I have only a small astrolabe. [...]

Galilei's answer has been lost, but Altobelli wrote a second letter shortly after:

(107) Ilario Altobelli to Galileo Galilei
Verona, 25 November 1604

I am very pleased with your reply, which makes me understand that you have consideration and affection for me. I am sure that the occasion of this wonder of the sky, [...] will lead us to know the dynamics and truths of the celestial nature, as never before in the past centuries.

It is impossible that the Star is a globe suspended in the air due to some kind of denser humid air, which is what the celestial fires feed on, because we see that it does not have any kind of motion as it should have given the liquid and mobile nature of the air. It is similar to the other stars of the eighth sphere: it has never changed color, it sparkles more than any other fixed star; its position makes it possible to imagine everything that Aristotle claimed to be impossible, as it is located in a southern part of the zodiac, close to the ecliptic; it was born in a fire sign and is close to the Sun. Therefore, it is undeniable its nature of burning flame. But if these Peripatetics, or better to say, semi-philosophers, persist in not wanting to understand the obvious demonstration of the fact that the Star resides among the fixed stars, and that it exceeds three hundred times the size of the Earth, how can we convince them of everything else? Galen,[5] in his *De diebus decretoriis*, asserts that it is dishonest not to want

---

[5] Galen of Pergamon (c. 129–201) was a Roman physician originally from Asia Minor. His views dominated Western medicine for thirteen centuries, until the Renaissance.

to learn and not to believe those who know, and that it is typical of pedantic and uselessly quibbling people to deny the evidence. I believe that the power of an erroneous education is too powerful, because only a blind indoctrination can cause such an unshakable obstinacy, to the point that not even the clearest truth is able to move anything. I am convinced that if Aristotle himself came back to life, he would change his mind. In any case, it will be the Star itself [...] to sweep away every wrong belief, and will be the guide so that we can finally walk following the light of reason.

I think I was one of the first, if not the first, to notice its first appearance here in Europe, that happened on October 9 almost at sunset, during the Conjunction of Jupiter and Mars.

The star, seen with the naked eye, seemed to have the same length as the two planets, and what we saw was something like this:[6]

Bor. ✳

Or.   ♃   ♄ **Oc.**

♂

Au.

But reading Your observations, I am convinced that my eye was mistaken, perhaps due to some refraction, and all the more so since the theologian P.D. Mordano sent me the precise measurements made by a disciple of Tycho Brahe with an instrument designed by Tycho himself, which in addition to confirming the absence of parallax and motion, provided observations very consistent with yours, namely, in Sagittarius, 17°51' with latitude 1°41'. From Augusta in Germany somebody sent me the value of 21° in Sagittarius; from Rome I received 14°, but it may have been observed with dials or inaccurate instruments made by carpenters. I will receive in a few days the observations made by Magini, and I will communicate them to you.

On the Star I have sketched eight chapters, but I do not have time for now to fix them because I am trying to study since I do not master well the astronomical concepts. [...] There is an entire chapter on the actual date of its first appearance: in those days I was monitoring the sky on the occasion of the conjunction of Jupiter and Mars, and the evening of October 8, at sunset, there were only the three planets, arranged in the shape of an isosceles triangle:

Bor.

Or.   ♃   ♄ **Occ.**

♂

A.

No other star was visible in the entire sky. To observe the sky with me there was also a Father to whom I had taught to recognize the conjunction of those three planets. On the evening of October 9 we saw the planets in the same configuration as the previous evening. [...]

If the Peripatetics would like to fill the gaps of their philosophical beliefs, there are two things to do: the first, is that they are really willing to listen and open their mind, the second is that You would show them with extremely simple and practical examples how parallax works. Then maybe they would realize and could change their minds, like the ancient philosophers and sophists who denied the sciences but in the end were convinced thanks to mathematical examples. But if they do not realize that those who do not know how to look beyond the

---

[6] The astronomical symbols represent respectively: ♂ Mars,   ♃ Jupiter,   ♄ Saturn.

boundaries of their own discipline end up losing contact with reality, how could they ever be convinced? [...]

I am sorry not to be in Padua in this period, and I could not listen to your lectures.

Christopher Clavius also writes to Galilei.

(109) Christopher Clavius to Galileo Galilei
Rome, 18 December 1604

Here we talk a lot about the New Star, found 17° in Sagittarius, with south latitude of about $1° \frac{1}{2}$'. If You made any remarks, I would be pleased to be notified. Magini wrote me that he observed it in the same position, and so they write also from Germany and Calabria.

Later Altobelli makes a clarification on the date of the first appearance,

(111) Ilario Altobelli to Galileo Galilei
Verona, 30 December 1604

Pirro Coluti, my countryman competent in the matter, informs me to have received a letter from the Illustrious Mr. Bardi, in which he says to have recorded the first apparition on September 27 and to have observed it for more evenings, which is completely false. In the days preceding October 9, I had my eye pointed just in that part of the sky, intent to observe the motion of Mars towards Jupiter, and I did not see any star. Only on October 9 I saw it for the first time, orange like an almost ripe orange, and its sight caused us great wonder.

Even a doctor and mathematician from Calabria writes that before October 9 it had not appeared, having been also with his eyes upward in the previous periods. I am therefore surprised by those who report 27th as the date.

Magini sent me a copy of the letter[7] that was sent to him by Father Clavius, who observed it from Rome with the right equipment, and reports that the star is always devoid of motion and equidistant from many fixed stars, so much so that he places it in the eighth sphere.

Later there is a significant exchange of letters between Galilei and Onofrio Castelli.[8]

(112) Onofrio Castelli to Galileo Galilei
Roma, 1 January 1605

Our bond is so strong that I can only wish you a Happy New Year; I also want to remind you to give me the grace to give me some information, and also two words of judgment about this new star. [...]

---

(113) Galileo Galilei to Onofrio Castelli
Padova, January 1605

I have been asked several times by Orazio Cornacchinii[9] to send to Your Excellency, who very much desired it, a copy of the three lectures I gave in public about the light that appeared

---

[7] *Carteggio inedito di Tycho Brahe, Johannes Kepler ecc. con Giovanni Antonio Magini*, from Archivio Malvezzi de' Medici in Bologna, edited by Antonio Favaro. Bologna, Nicola Zanichelli, 1886, p. 283.

[8] Born in Terni probably in 1580, studied mathematics in Padua, where he was a pupil of Galileo; later he devoted himself mainly to research and studies of hydrological character and more specifically fluvial.

[9] Professor of Botany at the University of Pisa.

in the sky about October 9, called *stella nova*. I have apologized to Orazio, because I realize how weak my speeches are, and therefore not worthy of ending up in your hands.

[...] I have postponed this publication, and I will postpone it for a few more days, because writing what has been the main objective of my lectures, that is to demonstrate that the position of the new star is located far beyond the lunar orbit, is in itself very easy, obvious and trivial, to the point that, in my opinion, such a topic does not even deserve to leave the table on which I am physically writing it. It is a subject that I had to deal with because of the young students and the multitude of people who wanted to understand the geometric demonstrations, despite the fact that they were trite and trite, as well as banal, exercises in astronomy.

Despite the risk of censorship, I decided to expose my thought not only about the position and motion of this light, but also about its composition and creation, and I think I have reached a conclusion without contradictions. Therefore I had to go extremely slowly, and wait for the star to become visible again in the East, in order to be able to observe again with extreme care possible changes in position, size and brightness; therefore, continuing to shift through the different hypotheses, I believe I have come to something that is more than a simple speculation.

Thus I have decided to turn the lectures into a part of a wider discourse that I am currently writing on this subject; while the publication is therefore postponed, I wanted to inform Your Excellency that I have not yet sent you my lectures, not for lack of attention or because I do not take into consideration your person, but for the reasons mentioned above. Having decided to transform my lectures into a book and to add the part about the composition and creation of the new star, I will need more time [...].

Galilei wrote that he had almost immediately reached a "conclusion without contradiction, something more than a simple speculation" in a complete description of the *stella nova*: its position, its motion, its composition and its origin. He had decided to present his thought even if he was well aware of the risk of censorship he would have run. And in fact, when he had stated in his public lectures that the star was located far beyond the Moon, had incurred the wrath of the Aristotelians. He therefore decided to be extremely careful in examining all possible hypotheses, and to wait before (eventually and possibly) publishing his conclusions.

The fact that the supernova had no observable parallax proved that it was located in the firmament, thus the falsity of the immutability of the heavens above the Moon claimed by Aristotle and his school. This also confirmed Tycho Brahe's observations of the 1572 supernova and of the comets. Of the supernova he had precisely measured its position relative to the stars of Cassiopeia for 16 months (so much remained visible) without finding any displacement, and thus concluding that it was located beyond the planets, and had no motions of its own as the planets have. He had also measured the trajectory of the great comet of 1577 and other smaller ones, finding that they too were beyond the Moon, and concluded in his *Mechanics* that comets follow an orbit that no celestial sphere would allow. The absence of observable parallax gives instead no information about the cosmological system or the motion of the Earth because it is too far away.

In his lectures Galileo had spoken of a light that appeared on October 10 at sunset, stating that during the conjunction of Mars and Jupiter in the presence of Saturn, which occurred on October 9 at about 5 am, "we did not see in that area any other star than the three mentioned". As Galilei himself reported later, he saw the new star for the first time with his own eyes on October 28. The data on the observations had not been collected by him. But he was the person called to put together the pieces of an extremely complex puzzle to provide an explanation as complete as possible.

To those who later criticized his absence during those first observations, as if to emphasize a lack of complete mastery of the subject, Galilei replied sarcastically that "it seems that there is nothing more serious for those who work in science than not to have been the first among all to have seen the new star, almost as if there were an imperative obligation to spend every moment of one's life with one's eyes on the sky, waiting for some new thing to appear".

Galilei reported a light that initially was weak and then, in a few days, became so intense as to "surpass all the stars, both fixed and mobile, with the exception of Venus alone, a most splendid and entirely sparkling light, to the point that in the vibration of its brightness it seems almost to be extinguished and then immediately rekindled; a light that surpasses in splendor that of all the fixed stars, including Sirius itself; similar, for the color of the light, to the golden splendor of Jupiter and the reddish color of Mars". He also recognized how attractive was the idea that such light was generated by the conjunction of Jupiter and Mars.

In some notes written together with the lecture notes Galilei speculated on the origin, cause and composition of the new star, and one of his first hypotheses was that the star could not have originated from an explosion. Daily tests led him to conclude that objects that ignite rapidly also tend to burn at the same rate, while the supernova had remained visible until February 1606, so such a flame could not have had an explosive cause. He had also hypothesized that the new star could actually be a kind of "luminous exhalation", described as similar to those seen in the haze that covers the streets of Venice. An exhalation that would have been invisible from close up, while it would have been perfectly visible, despite the great distance, only at night and would have appeared in the form of "luminous vapors". Another possibility to consider was for Galilei the model of an evaporation emanating from the Earth, abundant enough to build a new star, and he proposed the example of the sky when it is completely full of clouds.

After two months, Galilei himself realized the invalidity of these last conclusions, which he defined as "weak and unworthy".

The presence of a new celestial body situated among the fixed stars greatly excited the followers of the Aristotelian school, the Peripatetics, who professed a simple, perfect, ingenerable and incorruptible sky, devoid of changes. The evidence of facts was different, and so they looked for interpretations and explanations that could make compatible that novelty of the sky with a model that denied its very possibility.

Some of them claimed that, since the star nova was a new body, it could only be inside the lunar sphere, that is between the Earth and the Moon, where, according to Aristotelian dogma, matter was allowed to modify itself. But to support this hypothesis they necessarily had to deny the validity of the parallax method, which showed instead that the star was well beyond the Moon, just as Galilei had explained during his public lectures. Others denied that it was new, and argued rather that it had always been there, among the fixed stars, but not visible; others still said that its creation was a divine decision, and therefore there was no reason to investigate further.

When the star reappeared, it could be seen just before sunrise in the East. It had become much fainter, equaling Antares in brightness. Altobelli wrote to Galileo.

(114) Ilario Altobelli to Galileo Galilei
Verona, 10 January 1605

I read that nonsense of the *Discorso della Nuova Stella*, [...] but he who is the cause of his own ills, let him weep for himself, and one cannot say anything else but that his brain does not seem to work.

I, and those who were with me, saw the star on October 9, not before. However, in the previous days, while we were observing that part of the sky, as well as on the evening of October 8, we wondered how it was possible that we could not see any other stars but the three planets, Jupiter, Saturn and Mars. On the evening of October 9, however, along with those three there was also the new one, and we wondered in fact... What star is that? Yesterday evening it was not there!

I think it was as big as Jupiter, and the color was like an orange half yellow and half green, but it could also look like a mixture of yellow and green. Because of the bad weather, it was impossible for me to observe it again, until the evening of October 15, when it appeared to me much larger than Jupiter, and indeed I think that this was the time when its size was at its maximum, so much so that in the following days it seems to have decreased in size. Even if it was, it had decreased very little, because continuing to observe it in the following days, I continued to see it always bigger than Jupiter.

The same writes to Father Clavius a Mathematician from Calabria, who reports to have seen the star on October 9 and not before, and that he is sure of it because he was also with his eyes to the sky intent to observe that part of the sky until the evening of October 8 included. He writes that, as soon as it appeared, it had the size of Jupiter, and then became rapidly bigger. I have a copy of the letter that this mathematician from Calabria sent to Father Clavius, who then sent it to Magini, and Magini sent it to me. [...]

Regarding the position, I can't say anything because I don't have enough precise instruments, [...] but relatively to the observations I made with the quadrant, I can tell you that for many weeks I measured it as perfectly equidistant from the other fixed stars.

In the early months of 1605, after the reappearance of the star, we still find two notes of Galilei (reported in the *Opere*).

On February 3, 1605, the new star was aligned with Arcturus[10] and the left knee of Ophiuchus.

On February 4, in the morning, one hour before Sunrise, the new star was in the middle of the line joining Saturn and Jupiter, and extending northward to Venus.

The decrease in brightness of the *stella nova* continued throughout the summer and, when it was seen for the last time by Kepler in October 1605, it had reached the fifth magnitude. In the spring of the following year it had become invisible to the naked eye.

---

[10] Arcturus is the brightest star in the constellation of Bootes, and the second brightest (after Sirius) in the northern hemisphere.

References

1. Antonio Favaro, "La nuova stella dell'ottobre 1604", in *Galileo Galilei e lo studio di Padova,* Le Monnier, Firenze 1883
2. William R. Shea, "Galileo and the Supernova of 1604", in *1604-2004: Supernovae as Cosmological Lighthouses*, edited by M. Turatto, S. Benetti, L. Zampieri, W. Shea, Astronomical Society of the Pacific, San Francisco 2005
3. Alessandro De Angelis, *I diciotto anni migliori della mia vita*, Castelvecchi, Roma 2021

Chapter 3
Galilei Against the Aristotelians

The *Discorso* by Lorenzini

A month after Galilei's lectures, in January 1605, the otherwise unknown Antonio Lorenzini published with the publisher Pierpaolo Tozzi of Padua a booklet entitled *Discorso dell' ecc. sig. Antonio Lorenzini da Montepulciano intorno alla nuova stella (Discourse by the excellent Antonio Lorenzini da Montepulciano about the new star)*. Lorenzini attacked the mathematicians and astronomers who had placed the star outside the sphere of the elements, and therefore, more specifically, his work was a criticism of Galilei—even if Galilei's name is never explicitly mentioned.

He argued that if that star had really been placed at a distance so great as all other fixed stars, then it could not have been new, or newly generated, because generability brings with itself corruptibility, and corruptibility is not a property of matter located in the eighth sphere. In fact Aristotle had spoken of ungenerable and incorruptible heavens: therefore Lorenzini stated that "the new star is not a star, but a kind of meteor placed at a distance from the Earth almost equal to that of the Moon".

Faced with the undeniable evidence of the star's lack of parallax, and in order to justify Aristotle's concept of the immutability of the sky, Lorenzini elaborated a series of new geometric rules, so complicated and confusing that no academic took them into consideration. But the inspiring principle of his arguments was that a technique such as parallax could not be applied to celestial things: physics is not valid outside the Earth and the circumterrestrial medium.

In his work, Lorenzini also explained that the presence of a new body in the heavens would be so destabilizing for the heavens themselves that any type of celestial body would suddenly stop moving. He also argued that the matter of which the heavens were composed had nothing in common with the four elements air, water, earth and fire, and therefore could not have any of the characteristics necessary to create or destroy matter.

© The Author(s), under exclusive license to Springer Nature Switzerland AG 2024   19
A. De Angelis, *Galileo and the 1604 Supernova*,
SpringerBriefs in History of Science and Technology,
https://doi.org/10.1007/978-3-031-59486-1_3

Lorenzini concluded his work by making astrological predictions and forecasts, as he claimed that the new star would influence the seasons, the crops and the physical and moral health of all humankind.

Lorenzini's *Discourse* contained the arguments of a world based on inflexible first principles and not on observation, that world against which Galilei was now accustomed to clash. Behind Lorenzini's name it is not difficult to see the inspiration of Cesare Cremonini (1550–1631), professor of natural philosophy in Padua and rival and friend of his colleague Galileo [1].

A summary follows of Lorenzini's *Discourse,* largely taken from Drake [2].

In the first chapter Lorenzini states that the new star has given rise to discussions between philosophers and mathematicians, and that he will expose his reasons for siding with the former, promising a future book in Latin on celestial matters in general.

Chapter 2 exposes some arguments of mathematicians. They say that the new star is in the sky,[1] and it is a fixed star because it sparkles and does not move. According to them if it had been in the air (i.e. under the Moon) this fact would have been in contradiction with our senses: our sight would have been interrupted a few miles away. Moreover the star, having necessarily formed in the circumterrestrial medium, would have then moved from the horizon to directly upwards covering some miles; but everywhere, in Italy and elsewhere, it was seen in the same place, and nobody has ever seen it ascending in the sky.

This chapter explains that, since no change in the position of the star was observed through parallax (note that Lorenzini misspelled the word "parallax": he wrote in fact "paralapse" instead of "parallasse"; Galilei made fun of him for this, as we will see later), mathematicians have concluded that the star is not in the air, but in the sky. In a similar way they place the Milky Way in the heaven of fixed stars, and not in the air, as Aristotle said.

In Chap. 4, Lorenzini asks whether the mathematicians' conclusions can be reconciled with Aristotle's doctrine that nothing new is ever created in the heavens. Lorenzini challenges them to explain what was disposed in the creation of this new star, since no old star was missing. Furthermore, how could one part of the sky be destroyed by another part to create a new portion of the sky, when the sky consists of a fifth substance (the "fifth essence") that has no contraries that could destroy it? If then the new star were made of fire, the element Fire would pop up in the sky, and all stars would be altered by it. If other elements were present in the sky, their presence would blur our vision; therefore this hypothesis also contradicts the observations. Moreover, if we would add a single star in the sky, these would stop rotating, as it was demonstrated by the philosophers: in fact the force that moves the stars is exactly that necessary and sufficient to cause the uniform circular motion that is observed.

---

[1] Here we mean the sky in the Aristotelian sense: the substance of the last orbits of the Universe, in which the Moon, the Sun and the remaining planets are located. According to Aristotle the celestial bodies are formed by an incorruptible substance or "fifth essence", and consequently are not subject to the cycle of generation and corruption that instead characterizes the four elements (water, air, earth, fire) that make up the bodies under the Moon.

Chapter 5 begins like this:

> Mathematicians are not satisfied, and they adduce the publicly exposed opinion that it is bad judgment to abandon what the senses tell us through observations and to go in search of causes. But you gentle spirits, who come to contradict the ancient beliefs not out of obstinacy but for the sake of truth, do not boast of having an opinion that is superior in this matter! If we were close to the new star, there would be no difficulty in accepting your conclusions; but since opinion is uncertain in this matter, there would be no difficulty in accepting them.

> Our argument is derived from the principles of physical things known to the senses when they are at a suitable distance and is confirmed by philosophical induction. Your argument is also derived from the senses, but is based on measurements on incredibly distant objects. And if your conclusion is based on astronomy ours is based on physics, and partly on that science called optics, which is appreciable not only for its certainty (which it derives from its father, geometry), but also for the wonders it promises. Because it is really a remarkable thing and a wonderful prerogative of our intelligence to be able to measure distances and the magnitudes of distant objects; and just this delight, which induces you to believe in the possibility of extending the certainty of your mathematical principles to very distant objects, is the cause of your deception (let it be said in peace) in the application of those principles, because you depart too far from the mother physics, the origin of astronomy.

This passage precisely exposes the fundamental issue between Galilei's science and official (Aristotelian) philosophy. Unlike the previous chapters, in which Lorenzini had naively assigned mathematicians irrelevant arguments with respect to their position (as Galilei will point out in his reply), the writer in the statement just quoted frames the question perfectly: is it possible to extrapolate experimental techniques that prove true on Earth to unknown regimes? Moreover, the criticism is addressed to mathematicians rather than to the reader. The writer refers to certain opinions as "ours" and "yours", while Lorenzini was committed only to report the arguments of others. Next, the words "publicly exposed", in the original "divolgata", reasonably refers to Galilei, given that the *Discourse* was the first published lecture on the Padua topics, and that Galilei was the only person, not to say the only mathematician, to have given a public lecture *ex cathedra* on the new star. Probably no mathematician before this event, and certainly not "mathematicians" in general, had ever made public the view that it was wrong to abandon sensible evidence and turn to philosophical reason—consistently with Plato's doctrine. If there was a mathematician who had expressed the opinion that the senses were to be privileged over first principles it was Galilei. Drake is convinced that this clear and precise statement was written by Cremonini and was specifically directed against Galilei's conception of the method of physical science.

The fundamental question was this. Assuming that science had to start from sensory evidence derived from bodies of reasonable size and within reach for examination, how was it to proceed to formulate conclusions about other things, in this case extremely distant things? Galilei asserted that the mathematical rules of measurement verified in ordinary experience could be rigorously applied to inaccessible things, when such rules could be found; and for position and distance the mathematical rules existed; therefore any contradiction with other kinds of reasoning must be the fault of the latter. Cremonini, without doubting the certainty of mathematical principles, declared them of uncertain applicability to things beyond the reach of direct evidence. Here, he said, the nature of things as obtained by induction from sensory evidence is the appropriate basis of scientific extrapolation; if mathematical deduction indicated otherwise, this must be a sign that mathematics was not applicable. As far as the position is concerned, it is an intelligent, defensible, and highly orthodox Aristotelianism. The fact that Cremonini had in mind as physical principles confirmed by philosophical induction some extremely dubious notions about the nature of the four elements and the inalterability of the heavens does not matter. Those things had to be argued on their own merits (as Galileo might have agreed). The dispute in question in this passage was whether in anything, including the measurement of distance, mathematics could ever resolve any actual physical question. On this, Galileo was as clear-cut as Cremonini was in holding the opposite position; he claimed that it solved the position of the new star.

Chapter 5 of Lorenzini's speech continued by saying that mathematicians were wrong in saying that a star in the aerial region must appear at zenith after a short trip and cannot remain in view for long after its rise, because it should be near the Moon, which does not set quickly.

This is an adequate preface to the content of the two chapters with number 6, which, as we will see, must have been inspired by Cremonini. Towards the end of Chap. 5, the spelling "paralapse" is used again, showing the hand of Lorenzini, but then the style changes. At the end of the mathematical section of Chap. 6 there is a praise of the parallax method. It seems likely that Lorenzini added his own observations to the material provided by Cremonini.

The two chapters numbered 6 show signs of authorship by Cremonini rather than Lorenzini. For example, the word parallax is spelled correctly, and Averroes[2] is reproached for an argument against Aristotle and Alexander of Aphrodisia that would hardly have been known, let alone considered important, by anyone but a professor of philosophy. In these two chapters is woven a highly sophisticated (and sophistical) argument against the astronomical use of parallax that cannot reasonably be attributed to the same author who misspelled the word.

In the first of the two Chap. 6 the author concedes the fact that parallax is mathematically valid, but argues that it does not apply to planets; for example, our vision cannot penetrate to the true center of any celestial body. Great errors creep in when men try to apply parallax to objects that cannot be seen. It can be shown that a celestial body is closer than another only when it eclipses or occults the other. But even planets are too far apart to be able to tell which one would hide the other. Moreover, mathematicians act rashly when they try to apply parallax to the Moon (the idea is that it is too big for us to judge precisely where its center is). Averroes was therefore misguided when he placed the Milky Way in the sky after having made parallax observations in Spain and in Morocco. Even more reprehensible would have been if he had supported (in agreement with Alexander of Aphrodisia) that Aristotle would

---

[2] Ibn Rushd, Latinized as Averroes, was a Muslim Andalusian polymath and jurist who wrote about many subjects, including philosophy, theology, medicine, astronomy, physics, psychology, mathematics, law, and linguistics. The author of more than 100 books and treatises, his philosophical works include numerous commentaries on Aristotle, for which he was known in the western world as a strong proposer of rationalism and Aristotelism.

agree with the mathematicians. Natural philosophy would be ruined if we admitted into heaven mechanisms of creation and destruction that are appropriate only for our lower regions.

In the second Chap. 6 it is argued that the Moon must be seen in the same place among the fixed stars from any point on Earth. An alleged mathematical proof of the hypothesis that the errors of observation and calculation are necessarily greater than the parallax of the Moon (as Kepler sarcastically said, this would prove that the astronomers cannot be trusted) is shown. The long and sophisticated demonstration depends on a diagram that is missing in the book (perhaps Cremonini did not provide it and Lorenzini was not able to produce it). Lorenzini was so fascinated by the demonstration, based on the impossibility of seeing the whole circumference of the Moon or to know the precise location of its center, that he said it deserved the sacrifice of a herd of oxen, as it had happened, according to a legend, for the Pythagorean theorem—although to the Blessed Virgin and to Saint Anthony of Padua, not to the pagan gods. When the Discourse was reprinted, no correction was made either for the duplicated chapter numbers or for the inconsistent spelling of the word "parallax". This tends to confirm the existence of two separate authors.

There is no reason to think that Cremonini had anything to do with the remaining chapters.

Chapter 7 explains Aristotle's analysis of bodies located in the regions of air and fire, where vapors and exhalations cause optical effects that apparently place them in the heavens. Lorenzini denied that (as some moderns believed) air or fire could touch the circle of the Moon, because such a contact would cause great heat due to the rapid movement of that sky. The air consists of two regions; the heavy and cold vapors in the lower part cause rains and winds, while higher up there are fiery vapors such as rainbows, comets and meteors. The Milky Way is composed of lighter vapors that have made their way to the lunar sphere, where they are illuminated by starlight. Comets and new stars are composed of vapors separated from the Milky Way.

Chapter 8 reports quotes from the Neoplatonist Macrobius[3] and the Platonists who showed that the Milky Way moves with the Earth's seasons.

In Chap. 9, the author explains how it is possible for a part of the Milky Way to separate from it, and remain in a fixed position through an internal force and resistance to the attraction of the Galaxy itself. These parts form comets with or without tails, swept with all the velocity of the lunar sky when they are condensed by starlight and reduced to globular form. The new star of 1604 was influenced by Saturn, which is very fast and vigorous in action, as explained by Albertus Magnus[4] in relation to the formation of the human fetus. It remained without tail because

---

[3] Macrobius (c. 385–c. 430), a native of North Africa, was a Roman writer. A scholar of astronomy, he supported the geocentric theory. In his neo-Platonic thought God creates minds, which create the incorporeal soul of the world, which in turn degenerates to become the matrix of physical entities.

[4] Albert Magnus (1205–1280) was a Dominican, bishop, writer, and philosopher. Is considered the greatest German philosopher and theologian of the Middle Ages (he studied logic, physics, astronomy, biology, mineralogy, chemistry, as well as philosophical disciplines), and the greatest scholar of Aristotle. He was a teacher of Thomas Aquinas.

the Moon was very far away, being almost diametrically opposite at the time of its formation.

In Chap. 10 the twinkling of the star is explained as an effect that occurs in distant bodies because our vision is weakened by distance. Even though the new star was closer than the Moon, it still twinkled because it was made of a watery material that attracted some oily vapors, so that its fire spreads while being shaken quickly, like that of a flashlight (this does not happen with planets because they are made of more noble material). Pliny considered the dark spots on the Moon as dirt of the Earth ascended with the vapors. Others believed that they were only some less dense parts of the celestial matter (therefore unable to reflect the light as well); others still, that they were denser parts (which therefore prevented more the vision). What Plutarch said, that the Moon is like another Earth, can be interpreted not as truth but as poetry; Lorenzini believed that Dante agreed.

The remaining chapters (11 and 12) set forth astrological predictions based on the appearance of the new star. Some appearances in the sky (like comets) are good for the Earth, others bad; this one was good. It would have consumed the excessive humidity of the air (the *Discourse* was published during a drought, see the opening of the Dialogue that Galilei will write in response to Lorenzini). As for humankind, the star would have purified our bodies, senses and minds, and inclined men to pursue the truth, like the star that guided the Wise Kings. We should take up arms under the guidance of the Holy Catholic Church, mother of Italy and most of the world. Even though the star governed only material actions and not spiritual ones, it was a sign that the latter should govern the former.

Altobelli dismissed Lorenzini's work by commenting only "I have read that nonsense of the *Discourse of the New Star*. The author has only himself to blame, and nothing else can be said except that his brain does not seem to work."

The *Dialogo de Cecco di Ronchitti da Bruzene*

The response to Lorenzini's speech was rapid and original, but also very detailed. On February 28, 1605 was published in Padua, at the same publisher Tozzi who had published the book of Lorenzini, the *Dialogo de Cecco di Ronchitti da Bruzene in perpuosito de la stella nuova (Dialogue by Cecco di Ronchitti from Bruzene about the new star)*, a pseudonymous booklet in Paduan dialect (a local form of Venetian dialect) written according to Favaro and to many critics by Galilei together with his student and Benedictine monk Girolamo Spinelli (who had a deeper knowledge of the Paduan dialect). It is not known how many copies were published; certainly a second edition was printed the same year in Verona by the publisher Merlo [3], with minimal changes (not substantial but with some improvements from the point of view of readability) to the text and a substantial change in the appendix, which contained some "stanze d'autore ignoto contro Aristotele" ("stanzas by unknown author against Aristotle"), in Florentine vernacular, which were substantially modified.

The *Dialogo* was an explicit mockery of what had been published by Lorenzini; it defamed the entire Aristotelian dogma, which explains why the author(s) preferred to hide under the pseudonym of Cecco di Ronchitti. Cecco di Ronchitti from Brugine (a small village of peasants near Padua), an unknown self-proclaimed farmer and land surveyor, gave in the book many indications of being in reality an astronomer from Padua. The fact that the book was written in Paduan dialect instead of Latin, which was the noble language used to deal with relevant subjects, was a sign that the topic around which the booklet was built was not worthy of consideration.

In addition to the fact that the dialect was perfect for light-hearted teasing, the dialect of Padua was in vogue thanks to the writings of Angelo Beolco, a sort of *ante litteram* "blogger" who had published in the first half of the sixteenth century under the name of "Ruzante" or "Ruzzante", which in Paduan means "scratcher" or "mumbler", about the events that took place in Padua.

The Dialogue takes place between two peasants, Matthio (Matteo) and Nale (Natale), who confront each other after Natale read Lorenzini's book and summarizes to Matteo its contents. The booklet analyzes point by point the contradictions and distortions expressed by Lorenzini in his *Discourse.* It seems that Galilei did not want to react personally to what Lorenzini had written, but did so through the dialectal, intelligent and frank voices of two peasants.

It is easy to recognize in the text many characteristics of Galilei, so that already at the time of its publication the treatise was attributed to him. In addition to this, no family named Ronchitti is recorded in the birth registers of the parish of Brugine or in the neighboring towns; moreover, Cecco is also a Tuscan nickname. Galileo had good knowledge of the Paduan vernacular: Niccolò Gherardini, in his *Life of Galileo,* writes: "he was still very familiar with a book entitled *Il Ruzzante,* written in the rustic Paduan language, taking pleasure in those crude tales and ridiculous incidents" (Gherardini confuses the author with the title). Among the *Inedita Galilaeiana,* published by Favaro in Venice in 1880, one can read a thought of Galileo that contains words of the rustic Paduan dialect. Finally, in his letter to Paolo Gualdo of June 16, 1612, published in volume XI of the *Opere,* Galileo writes whole sentences using the same dialect, and Galileo's Venetian correspondents often wrote to him in various flavors of the Venetian dialect. In a letter to Paolo Gualdo dated August 16, 1614, Girolamo Spinelli was called by Galilei "one of my students, monk of Santa Giustina, companion of Cecco de' Ronchetti".

In the mouth of one of the two interlocutors, Matteo, there are Galilean concepts; we can say that Galilei's ideas about the new star are all in Matteo's words. The Copernican idea of the author of the *Dialogue* is in the explanation of the concept of parallax made in the booklet, which also contemplates the seasonal displacement of the Earth linked to its revolution around the Sun. Two drawings reproduced from Galilean manuscripts kept at the Biblioteca Nazionale Centrale di Firenze, presumably contemporary to the publication of the *Dialogue,* and included in a section entitled by Galileo "De stella anni 1604", clarify even better the meaning of parallax applied to the Earth's revolution.[5]

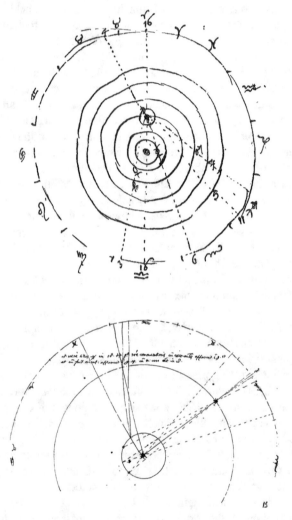

Note the Copernican essence of the drawings, and the outer circle of the Zodiac with the signs. In the first drawing the Earth, third planet from the Sun which is the black point in the center, is indicated by a sign badly written, as if Galilei was afraid that his notes would fall into the wrong hands, but it is still between Mars and Venus and is the center of the observations—a drawing similar will be reproduced in the *Dialogue concerning the two chief world systems* in 1632. In the second you can see

---

[5] Manoscritti galileiani 70, c. 16v e 47, c. 10v. Immagini riprodotte su concessione del Ministero della Cultura, Biblioteca Nazionale Centrale di Firenze.

the seasonal variation of parallax. The drawings are not in the original first edition of the *Opere*, but were added in an appendix to the second volume in later editions.

The *Dialogue*... begins with Natale talking about the land being particularly arid and dry at that time, and informs Matteo that a "doctor from Padua" had attributed the drought to the appearance of the new star. The two peasants also wonder if the new star is really new or old, and why someone would think it as old if nobody has ever seen it before 8 October. They then begin to speculate about the position of the star: Matteo, who has not read the book but knows that Lorenzini is not very prepared or sure about mathematics, wonders if it makes sense to take into account the considerations of a philosopher who does not master mathematical knowledge and who, despite this, decides to deal with subjects that do not belong to him, or if it is more appropriate to rely on mathematicians and astronomers. When Natale mentions the attacks that Lorenzini had made against mathematicians, in which he accused them of necessarily having made mistakes in their measurements (since they insisted on placing the star in that part of the sky that does not allow the appearance of new matter), Matteo replies that mathematicians should not be interested in what others think is possible or not, but only in measuring and drawing conclusions from their measurements. The two peasants discuss many other arguments, such as the Aristotelian hypothesis that the presence of a new star would have stopped the motion of any celestial body, because its appearance would have destroyed a well-ordered harmony, as Lorenzini pointed out. Finally, they remind that the concept of an immobile Universe has already been considered possible by many illustrious scientists, but no possibility can dispel one of the most obvious observations: the star was not there before.

In line with what Favaro did in his *Opere,* I report the full translation of the *Dialogo*... for which I have taken as a starting point the translation by Drake [2] who in turn started from Favaro's version. I have largely used the edition by Milani [4], and a little that by Bozzolato [5].

In addition to being more compelling than a treatise, as Plato had already shown and as prescribed in numerous courses on rhetoric of Galilei's times, a dialogue allows for circumventing formalism in certain conclusions that Galilei was probably not able to develop rigorously: a dialogue allows its participants to forego certain rigorous demonstrations and replace them with assumptions of sufficient plausibility, made stronger by the use of common sense and of humour. In this sense the technique of persuasion in this dialogue anticipates by three decades the two great Galilei Dialogues: the *Dialogo sopra i due massimi sistemi...* (1632) and the *Discorsi e dimostrazioni matematiche...* (1638). A focus on the form of the text of the *Dialogo* caused however Galileo's contribution to be greatly diminished [6] in editions following Favaro's (especially from Paduan critics), in favor of Girolamo Spinelli. This gradual desauthorization has certainly also depended on excessive attention to form, admittedly so particular, which has unfortunately come at the expense of attention to specific content, which is equally (if not even more) important [7].

In addition to the *Dialogue* I reproduce with a minimal transliteration the poem in octaves "by an unknown author" which appears in the appendix to the first edition. The poem seems to me attributable to Galilei (who was also skilled in the art of poetry), also in consideration of the style and of the learned astronomical considerations that appear there. It does not appear in this form in later editions. It does not appear even in the national edition, and some (see [8], also for notes on the literary production by Galileo) possibly attribute it to Antonio Querengo,[6] to whom the Dialogue is dedicated.

The following objective elements strengthen the attribution of the poem to Galilei:

- certainly Galilei had a strong participation to the *Dialogo de Cecco di Ronchitti*; being a poet (the full volume IX of the National edition of his *Opere*, edited by Favaro, is dedicated to his literary activity), it appears strange that the poem in the Appendix of the *Dialogo* itself was written by somebody else, or at least that Galilei did not heavily rework it;
- the style, including the verbal violence, is coherent with Galilei, and there is a quite a strong attack to the Aristotelians;
- as Milani observes in her 1992 translation of the *Dialogo* [4], some rhymes are common with those of Galilei, and one stanza begins with the imperative "Dunque", as in Galilei's reworking of Andrea Salvadori's *Canzone per le Stelle Medicee* (vol. IX of Galilei's *Opere*);
- a description of the method of parallax is appropriate for an astronomer, and a Copernican statement is present in the fourth *stanza*.

Even if the poem had not been written entirely by Galilei, I am still convinced that there is enough evidence to show that Galilei had copiously reworked it.

A second edition of the *Dialogo de Cecco di Ronchitti* was published later the same year in Verona [3]. The poem was largely amended, with some uncertainties in the poetic language. In addition, the dangerous Copernican allusion to Earth rotation in the fourth stanza was removed, a stanza was added, and some criticisms to Aristotle were softened and became "politically correct" (for example, in the first verse the sentence "stolto Stagirita", where "stolto" literally means "fool", "foolish", or even "stupid" [9], became "per altro Stagirista saggio", i.e., "otherwise sage Stagirite"). This new version, compared to other poems by Galileo, does not seem to come from the same pen, and lacks Galileo's verbal aggressivity.

The references made to Lorenzini in the first edition of the *Dialogo*, marked in the text by notes in the margin "[Lorenz. Cap. x]", are given below as endnotes to the chapter, numbered with Roman indices and a translation of the original text to which they refer is provided. The footnotes are mine, and are numbered with Arabic indices.

---

[6] The Paduan Antonio Querengo (or Querenghi, or Quarenghi, or Antuogno Squerengo in the dialectal mispronunciation) was a very learned diplomat and poet in Latin, a friend of Torquato Tasso. After the death in 1601 of Giovanni Vincenzo Pinelli, a good friend and sponsor of Galilei since his arrival in Padua, Querengo's house replaced Pinelli's as a mecca for all important visitors and a meeting place for local scholars and men of letters. Among these were Paolo Gualdo and Lorenzo Pignoria, who were also good friends of Galileo. Querengo continued to follow with interest the career of Galileo after he left Padua, and it is from his pen that we have the most amusing letters concerning the unfortunate campaign of Galileo in Rome in 1615–16 to support the Copernican point of view.

DIALOGO

DE CECCO DI RONCHITTI

Da Bruzene.

IN PERPUOSITO

De La Stella Nuova.

Al Loftrio e Rebelendo Segnor Antuogno
Squerengo degnetiffemo Calonego de
Paua , sò Paròn.

*Con alcune ottaue d'Incerto, per la medefima Stella,
contra Ariftotele .*

In Padova,
Appreffo Pietro Paulo Tozzi. M.DC.V.

To my Illustrious and Reverend Patron Signor Antonio Querengo Most Worthy Canon of Padua

What would you say, Reverend Patron, if you saw a poor servant of yours, never occupied with anything but herds[7] and having no business except to survey the fields, in a quarrel and dispute with a Doctor of Padua? Wouldn't that make you laugh? However, a cancer might take me, that's how things stand. Only see how I managed this like that fellow that dressed up in somebody else's clothes so as to look like a Doctor.

Fact is, ever since I was a boy I've been a nosey fellow. I always did like to gaze at Venus, Orion's Belt, the Bullseye, the Seven Sisters, and the Big Dipper. But I could never have reasoned about them if I hadn't heard you often talking about such things. Like this New Star that's got everybody puzzled—you have sure discussed about it, arguing with anybody that says it wasn't way up in the sky. Not that I was one of them; but I did push up to the edge of the crowd and hang around listening, and though I'm no highbrow like them, I've thought a lot about what you said.

So now that I've put all this down on these worthless pages, you see how I put your coat on. If I make a good show, the credit is yours. But if by bad luck (as I don't think) there should be a piece of shit that strips me, then it will be up to you to help me, since it's yours. I do hope you like it, dear Master, and also I pray God for your long life and good health.

Padua, the last day of February 1605
Your servant,
Cecco di Ronchitti

———————————

Dialogue between Matteo e Natale.

Matteo. My God, what a drought! What a roasting! I just don't know what to think. Good-bye water! They say at Venice that the lagoon is drying up inch by inch. Just imagine, at Lizza Fusina[8] one can walk. People only hope the grain will sprout. Oh, it'll come up, as the man said.

———————————

[7] The word *boaría* (herds) used in the original refers in Paduan also to the behavior typical of primitive and vulgar people.

[8] A marshy area near Venice.

Natale. Hello, Matteo! What on Earth are you mumbling about? Are you talking to yourself?

Matteo. Oh, hello, Natale. Brother, I don't know myself. I was just racking my brain wondering why it doesn't rain. How about you, is it dry enough for you? Is there any danger that the river banks will let go under a flood?

Natale. As to that, it's said that whenever a rain cloud shows up it goes away again without wetting the sand as much as a pissing frog. I think that if this continues, it will be the end of the world. The meadows are all dry and full of cracks, the countryside hardened like a bone; so that in the long run I expect a hard time for us and for the cattle.

Matteo. Pull yourself under this walnut tree for a while. It's more than an hour till evening, anyhow. Do you have any idea where this drought is coming from?

Natale. Didn't you see that star three months ago, shining at night like an owl's eye? The one that you see now in the morning when you go out to prune the grapevines? The very bright one. Well, it was just born then and had never been seen before. That's what's really causing these freaks and this drought, according to what a Doctor from Padua said [Lorenz. Chap. 2][9]

Matteo. How do you know it was never seen before?

Natale. The other day I heard a man that was reading a little book, and he said it only began to show last October eighth. The book was written by a professor from Padua, and said a lot of things.

Matteo. Might a cancer come to him and these to those little shits of Padua! Maybe just because he never saw it before, he wants everybody to believe that it wasn't there. Me, I've never been to Germany, but yet it is there.

Natale. Well, now, I don't know; me, I thought it was new, too.

Matteo. I don't say it wasn't. But the thing is, that kind of reasoning wrong, even if it did come from a schoolteacher.

Natale. All the same, we've got to admit it was new.

Matteo. Sure, but so far away he can't know what the hell it is.

Natale. Far? Nuts. It's not even as far as the Moon, from what the book said [Lorenz. Chap. 5].[10]

Matteo. What is this fellow that wrote the book? Is he a land surveyor?

Natale. No, he is a Philosopher.

Matteo. A Philosopher? What has philosophy got to do with measuring? Don't you know that a cobbler mustn't talk about buckles? It's the Mathematicians you've got to believe. You have to believe the mathematicians, who are measures of the heavens; just as I measure the countryside and can rightly say how long and wide it is, so do they with the stars.

---

[9] [Lorenz. Chap. 2] "One of those never seen before, as is evident from being in addition to the number of stars already diligently determined" (*Discourse...* , p. 4v).—"And outside of this star, no one could point to any other cause of drought, poor wind, and damage to the abundance and goodness of crops and animal parts" (p. 29v).

[10] [Lorenz. Chap. 5] "That it should not be understood to be far from the Earth, but close to it, and almost contiguous to the lunar orbit. But what the distance is it is impossible to determine" (p. 10v).

Natale. He said there in that book that the Mathematicians think it's very far out, but they don't understand.[11]

Matteo. How don't they understand? Is he joking or is he serious?

Natale. He says they imagine that the sky can be destroyed or created a bit at a time, though not all at once. How should I know.[12]

Matteo. Now, where do Mathematicians talk that kind of reasons? If they just stick to measuring, what do they care whether or not something can be created? If it was made of polenta, couldn't they still see it all right? That wouldn't make it any bigger or smaller, would it? These boners of his make me laugh.

Natale. The thing is that he writes these things all through the book[13]

Matteo. Well, he is unexperienced... Let him get out of this as best he can.

Natale. He says if it was just created in the sky, then another star or something else must have been destroyed where it was, or nearby, to make up for that. But nothing is seen missing[14]

Matteo. I hope you don't think that I want to reason like a Mathematician. But this thing is so nonsense that I cannot keep silent. Let's say that right here is a little bit of sky, and over here is some more, and we put them together. Can he say where any is missing? When clouds gather, or it rains, what sign is there that something was taken to put them together? And now for the star. Since he wants it created in the air, where was the air ever thinned out? Also, if he imagines that all the stars there are in the sky can be seen, then somebody ought to tell the priest about him. Besides, who is going to stop me from saying that maybe three or four smaller stars that couldn't be seen, or even more, were heaped together to make this fine large one? Or maybe it was started in the air and then always went higher up... I don't claim such things are so, because it isn't my business and I don't know the first thing about it. It's enough for me that he doesn't reason any better.

---

[11] [Lorenz. Chap. 2] "The sages are all occupied in this doubt, whether it is placed in the starry sky, as the mathematicians say, or in the air above the Earth, as the philosophers say" (p. 4).

[12] [Lorenz. Chap. 4] "If you think that this new star, as mentioned above, appears in addition to the number of the others, it is certain that it was generated as new, since it cannot be said that it was hidden, clear and large as it is, in the perfectly transparent sky. Nor can we understand why and in what way it gradually came to show itself. But surely this appearance and this increase are signs of a generated thing, from which it is clear that it cannot exist in the heavens, since they, although created by the Author of nature, who could, if he wanted, annihilate them, are ingenerable and incorruptible, as Aristotle shows. Even if mathematicians do not agree, and say that the sky can be corrupted?" (p. 6v).

[13] [Lorenz. Chaps. 4–5] "Therefore it must be known that generation is joined to corruption: the generation of one is the corruption of another" (p. 7r). "Therefore the mathematicians do not rest, and claim that it is weakness of judgment to leave the sense and search for the reason. But, O gentle souls, who not out of obstinacy but for the sake of truth contradict him, do not boast of having manifested this sensation in this matter, for if we were close to the aforesaid star, there would be no difficulty, but since there is no certainty about the sense of things so distant, you too proceed with sensations (p. 9v).

[14] [Lorenz. Chap. 4] "As Aristotle says, in all the past time, according to the traditions of the ancients nothing appears changed in the sky, not even in part. Therefore it would be astonishing that a new star should now be generated or corrupted without generating or corrupting others" (p. 7r).

Natale. But he wants this to be the backbone of Aristotle's reasoning.[15]

Matteo. With such weak sinews, all his stuff about creation and destruction will become a soup.

Natale. If the sinews are weak, the meat sure will be tender. He says that if new stars could be created in the sky, then in past times some would have been destroyed that have always been seen, and that comes to... I don't remember how many.[16] Anyway there were a lot, and not a single one is missing, and that's just what Aristotle said [Lorenz. Chap. 4].[17]

Matteo. Very weak... Who the devil told him that this new star is a starry star? Maybe it's just a bright spot with no star. I didn't call it a star till now, because it isn't one; it just looks like those others.

Natale. Well, what is it, then?

Matteo. How should I know? It's not a real star, and that's enough. Other stars were not destroyed, because they are stars and the sky has need of their doings. But it doesn't need this one, and because this came, it will also have to go. As to saying no star was ever seen to be destroyed, let me ask you something. The Earth is not as big as a star, but was it ever changed all at once?

Natale. Great balls of fire! If the Earth had ever changed that way, wouldn't everything have got messed up?

Matteo. That's what I think. Things only happen a little at a time, and that's how it must have been with stars that really are stars. Next I'd like to ask this fellow with the book how he knows no star was ever lost bit by bit? As to his saying that nobody ever added any" and that Aristotle said so, that's news to me.

Natale. He says that if this new star was in the sky, all natural philosophy would be nuts, and that Aristotle thought that with one star added the sky could not move at all.[18]

Matteo. Cancer on this star, then; it did wrong to ruin the philosophy of those fellows that way. If I were them I'd have it hailed before the Mayor and charged with illegal possession. I'd put up a tough fight, I would, and get a public and personal order against it for causing the sky not to move. Yet that's not so bad, either, because there's a lot of people, and good ones too, that believe the sky doesn't move.[19,20]

---

[15] [Lorenz. Chap. 4] "Therefore I propose to you the backbone of Aristotelian reason, which, according to that philosopher, is not so completely flattened" (p. 6v).

[16] Lorenzini gave the number as 1022, mentioning also Pliny's figure of 1600.

[17] [Lorenz. Chap. 4] "Since in the past days that star has been generated, it seems rational to explain why in so many thousands of years none of the 1022, or 1600 according to Pliny, stars of the Firmament has been corrupted, since if a new star is generated it seems reasonable to me that somewhere another one disappears" (p. 7).

[18] [Lorenz. Chaps. 4–5] "Rightly the Philosopher wanted to say, that if a new star was added to the Sky this one could not move anymore" (p. 9). [Editor's note: Aristotle's argument is that the firmament has the "right" number of stars to make it move with uniform motion around the Earth].

[19] [Copernicus, etc. 4].

[20] In the next edition the sentence was changed to read: "This could also be put down by telling the lie that the sky did not move even before this star was born, and citing as false witnesses those of

Natale. Now hear this. He asks why that higher sky should be less complete than the others, saying it would be less perfect and corruptible giving birth to new stars, which doesn't happen in the lower ones[21]

Matteo. Cancer... That sky is as much bigger than the others as Mount Rua[22] is bigger than a millet seed. And being so big it may have more new stars, not like those others that have enough with one apiece. Still, if some little starlet was also born in one of those, does he think everybody would see it right away? Boy, what a chump![23]

Natale. He says the world can't be complete without something that's free from creation and destruction, which must be the sky.[24,25]

Matteo. The sky? Why the sky, exactly? I say that it is Paradise, above the sky, that is pure the way the Doctor wants.

Natale. He thinks it couldn't be that such a giant star should jump out all at once, like a stroke of lightning.[26]

Matteo. Well, when a cow calves, the moment the calf is born it is bigger than a fully grown lamb. Why? Because the calf's mother is a lot bigger than a sheep. And mind you that this star, in the whole sky, is smaller than a lion or an elephant is in the whole Earth. So is there any great wonder here, as you see things?

Natale. Well, if that's how it is, why is the star shrinking instead of growing?

Matteo. If you ask me, I think it keeps on going up, and so it seems to dwindle because it's getting farther away.

Natale. Easy, now. The book says it was seen to grow at first, and quite a bit. If it was going up and couldn't stop, why wouldn't it always have been shrinking?[27]

Matteo. That fellow with the book must have needed glasses. Me, I know it looked pretty big to me the first time I saw it, and ever since then it has lost size, so to speak. Anyway the arguments don't convince me, and that fellow with his book is getting lost, so I'd like to get him back on the road.

---

the Pitariegi [literally "gobblers"] and of the Coverchi de Cane ["Covers-of-cans] who want things that way." A note was added "Mentions the Pythagoreans and the Copernicans."

[21] [Lorenz. Chap. 4] "These views would imply the imperfection of the heavens, which instead were created perfect." (p. 9).

[22] A hill in the country near Padua.

[23] In the original *cottora,* in Paduan dialect "fava bean". The word does have a meaning in Tuscan, not in Paduan.

[24] Cf. *De caelo* 270b, 14. Aristotle regarded novae as a special kind of comets in the region of fire below the Moon: *Meteorologica* I, vii.

[25] [Lorenz. Chap. 4] "And in order that nothing be silent, it is shown that the same sky would be contrary to itself, because within its diaphanous no star is generated or corrupted, and so much is sufficient to prove that the heavens are incorruptible and ingenerable." (pp. 8v and 9r).

[26] [Lorenz. Chap. 4] "It is astonishing how it could be generated in such a short time."

[27] [Lorenz. Chap. 1] "At first it was seen to be small, and then growing from day to day it became in appearance of size and light not inferior to Venus, and superior to Jupiter, besides all the fixed stars, to which it resembles in twinkling. It is fixed in its site and as the fixed stars has a movement every 24 hours from east to west" (p. 3v).

Natale. All right, listen to this. He says nothing can be created in the sky, because—he really says this—contrary things would have to be up[28] there, which they can't be, since the sky is a fifth sum, or substance,[29] I forgot which.

Matteo. Oh, yes. Onions![30] These are some of those keen words of Aristotle and his friends who don't even know if they're mortal and yet want to reason about the sky. I think that in the sky there are hot and cold, wet and dry, just like down here. Thick and thin are seen there, and so are light and dark. How about that? Those are contrary things. What more do you want? That star could have been there, but it wasn't, and now all of a sudden it is. Isn't that a switch? This fellow just opens his mouth and lets out anything that comes. He may want to tangle with Mathematicians, airing his arguments, but where did he ever find a measurer worrying over such make believe? Who told him they do?

Natale. A cancer on this. But he goes on to say that if Earth and air and water and fire were in the sky, we couldn't see through it the way we do, because it would be thick and dark [Lorenz. Chap. 4].

Matteo. Sure, if those elements were the same as ours. But they are more perfect, my Master once told me, and he said that Plato said so.[31]

Natale. Also he says that in that case the sky couldn't go around, because the elements go up or down, not around.[32]

Matteo. What if I contradict him and say they also go around? He leaves out writers who say that the Earth goes around like a mill. And you've got to think about other people, because when it comes to arguing, everybody can participate.

Natale. Next he says the star is near the Moon, but under it, so it can't be anything but fire.[33]

Matteo. Better he might say that there isn't any such fire, for a lot of reasons.

---

[28] [Lorenz. Chap. 4] "It is manifest that generation and corruption occur among the contrary, and that the change of forms has an agent. But heaven is a fifth substance, and has no contrary, therefore in it there can be neither generation nor corruption. Mathematicians, on the other hand, admit that there is contrariety in the heavens, and therefore it is convenient to remove all doubt from their minds. It is evident that contraries are bodies that have opposite qualities with respect to each other, that is hot or cold, wet or dry. Therefore Aristotle collected the number of the four elements. Therefore, if there is contrariety in the sky, one concludes that it is either Earth, or Water, or Air, or Fire, and they agree with Plato (though not understanding what Plato means about immaterial fire) that the sky is Fire" (pp. 7v–8r).

[29] Lorenzini wrongly wrote "quinta sostanza" ("fifth substance") instead of the proper "quint'essenza", meaning Aristotle's quintessenza as distinguished from the four elements.

[30] The word "onions" may refer to the skies, thought of as nested like layers of an onion, a metaphor already used by Galilei; or it may just mean "Balls!"

[31] [Lorenz. Chap. 4] "I do not want to reason with those who say that among the heavens are the elementary matters, because this contradicts the order of nature and common sense. It would be against their nature, and then the Earth, mixing with the sky, would make it opaque" (p. 9r).

[32] [Lorenz. Chap. 4] "But that it is not fire, and similarly, that it is not any of the four elements, is evident, because they do not have a circular movement, but straight, upward or downward, which is not found in the heavens" (p. 8r).

[33] [Lorenz. Chap. 7] "The said star cannot be other than the Moon's sky" (p. 19r).

Natale. So he holds that air licks the Moon's ass— I mean, its sky [Lorenz. Chap. 7].[34]

Matteo. Ho ho. And well he may say that, too.

Natale. But, he says, the sky can't be fire, because that big a fire would[35] burn up everything else.

Matteo. Plague take me if that fellow, Doctor though he is, wouldn't look the same as anybody else with his clothes off. Listen. Wouldn't a tiny spark kindle a whole haystack, and burn any amount of wood there was?

Natale. Yes, I believe it would.

Matteo. And yet all the furnaces in the world can't burn a ducat if it is pure gold. Why is that? You want to know why? Because gold won't burn; that's why. So if the other elements could be burned at all, it would take only a little fire to do the job. But if they can't, then everything would not burn up the way he said.

Natale. This fact demonstrates by itself. But do you believe the sky is really fiery?

Matteo. Who, me? I don't say that. It's just that this Doctor yells "wolf!" for no reason, and then says dumb things.

Natale. Hold on; listen to this one, since I couldn't care less about the[36] other. He says that the Mathematicians have good instruments and solid arguments but just don't know how to use them.

Matteo. Oh, how would he know? Is he the brother of the tower of the Bo?[37] All right, suppose a mathematician came and said to you, "Natale, I'm going to tell you how much air there is from this walnut tree to the riverbank," and then used his gadgets to measure this but without moving. After he had measured this and told you the answer, suppose you also measured it, with a string or something, and found he was right. Wouldn't you say he had used his gadgets well?

Natale. Yes, I would, but what are you getting at?

Matteo. Then when he measures to a star (so to speak), why should we say he doesn't know how to do that? And that if he is wrong, he is out by thousands and millions of miles? If the Doctor said "a little bit out," like four inches or a foot, I'd shut up, but not for this. This is too much.[38]

---

[34] "One can see the falsity of the opinion of the moderns, who say that it touches and licks the sky of the Moon" (p. 19r).

[35] [Lorenz. Chap. 4] "Fire being the most ample, which would burn the rest" (p. 8r).

[36] [Lorenz. Chap. 5] "That science which is called optics, so delightful not only because of its certainty, which it derives from its father, geometry, but also because of the wonderful things it promises, such as knowing distances, and magnitudes, and places seen from afar, is a singular thing. In spite of the certainty of your principles, however, [you mathematicians] deceive yourselves in applying them, departing too much from the mother of astronomy which is physics" (p. 10r).

[37] The central building of the University of Padua.

[38] These numbers are not in the Lorenzini book and the argument of Matteo is not based on what was said by Natale. Presumably Cremonini had said, in oral debate, that any astronomical distance derived from parallax must either be exactly correct or must entail some enormous error. That was implied, in a way, by the reasoning in Chap. 6 part 2, concerning which Kepler remarked that 52.5' of arc would have to be ignored by astronomers.

Natale. How do you know what reasoning by the Mathematicians he is going to count?

Matteo. Go ahead.

Natale. One, he says, is to cut away a little piece of a circle, and that way we could not see the star for more than half an hour. The other is to get right under it by walking miles in its direction.[39] He says they miss the point, just showing that the star is more than ten miles up, since even he says it is a lot higher than that[40,41]

Matteo. Cancer, this fellow is sharp on the thick side! But if he believes the star is more than ten miles up, and shows that those arguments have nothing to do with him, then why put them in his book and say they miss the point? From what was being said the other day at Padua, those arguments were made against a great Aristotelian Philosopher who did, at that time, claim hard and long that the star was no more than ten miles high. This fellow with the book should leave these arguments alone, since they give him no trouble.

Natale. Well, as the man that was castrating pigs said, is that all? Snip, snip, there's more! Next there's this awful mess about parallax, and optics, and the Moon, and what not. He explained these ideas three times over, and still nobody understood them.

Matteo. The Doctor must have tangled them up on purpose to make himself look smarter.[42] But that is going to backfire, because I know he must have gone wrong on parallax, which is a very sure way of measuring through air.

Natale. Let's see if I can remember anything about that. He says first that you can't see through a star, and then that since it is so far away you can't find its center, because it is round... [Lorenz. Chap. 6]

Matteo. Wait a minute, don't say too many things all at once. Who ever did think you could look through a star that is so thick? What rigamarole is he cooking up? He must know better than that. That's the first thing. Then as to the other—well, which is better for finding the center of a sieve: to put it close to your eye, or some ways off?

Natale. Why, standing away from it. Close up, I couldn't even see all of it [Witelo,[43] bks. 1&4, etc.; Euclid, Optics 13].

---

[39] [Lorenz. Chap. 2] "Supposing that a circle concentric to the Earth and very close to it, let us say ten miles high, is cut off from the horizon for a small portion of it, it is obvious that the new star being so close to us, would be above the horizon a few hours, and because of the roundness of the Earth would be hidden. Whoever walked towards its latitude, moving away from the Pole for a few miles without going below Sagittarius, would find it at the zenith" (p. 4v); "it is not to be understood as far from the Earth, but close, and almost contiguous to the lunar orb, whose distance is impossible to measure" (p. 10rv).

[40] In Chap. 2, Lorenzini had said that if the star were only ten miles up, it could not be seen by men beyond our horizon. A distance of 22 miles had probably been introduced orally by the "great Aristotelian" in debates at the University.

[41] [Lorenz. Chap. 5] "things unbelievably far away" (p. 10r).

[42] See the summary above of Lorenzini, Chap. 6.

[43] Witelo was a Polish writer on optics in the 13th century.

Matteo. Then why say you can't find the center of stars because they are far off? Next, think of a ball and a plow; which shows you its center best at a glance?

Natale. Cancer on it, the ball, because where you cut it in half at one point, it is so cut for all points.

Matteo. Yet he says the contrary.[44]

Natale. He goes on saying there that to our sight the stars are too tiny for the center to be found.[45]

Matteo. Oh yeah? He really says that, too? Tell me something; how could you make a bigger mistake, in finding the center of a barrel-end, or of a plate? To mark it, I mean.

Natale. I get you. I could be a lot farther wrong about the barrel-end than the plate.

Matteo. Yet the good Doctor of the book says not. Well, get on with that parallax.

Natale. He says that alas, being unable to see through the stars, you can't know just where they are because you can't see behind them.[46]

Matteo. If you hung your jacket on a step of my ladder so that the step was completely hidden, could you find out for me which one the jacket was hanging on?

Natale. You don't require much. I'd start counting one, two, three, up to where the jacket was. When I had counted say nine, and saw the jacket next, I'd say it was on the tenth. That's what I would do; isn't that right?

Matteo. That's the only way it can be for these, and that's how it's done in the sky, though this learned dwarf can't catch on to this. Oh, well, some people are as thick as the great tower of Cremona, which they say is gigantic.[47]

Natale. When we look at the Moon, he says, vision sticks through it (so he says) so we can't know its parallax [Lorenz. Chap. 6].

Matteo. Oh, he can stick his... (I nearly said it.) In that case we'd see the stars beyond the Moon, which would be fun.

Natale. Easy, now; I didn't mean to change his thought. It seems to me he meant the center of the Moon can't be seen, nor those of the stars, because they are wide and our vision is narrow, [Lorenz. Chap. 6] though it goes out widening.

Matteo. For God's sake be careful; whichever side you take, you're stuck. Why should I care if I can't see the whole Moon, or even a whole star? I see a bit and measure by that; isn't that enough [Lorenz. Chap. 6]?

---

[44] [Lorenz. Chap. 6] "We grant that parallax is true according to mathematical reason; nevertheless according to physical reason we affirm that it cannot be known in the planets, [...] since finding the center, so far away, is impossible, particularly in bodies spherical. [...] The intersections outside the center of the Moon in various parts of it form different angles; therefore, since the comparison is not made under the same condition, the parallax could not work" (p. 11r).

[45] [Lorenz. Chap. 6] "And even if the center of the Moon could be found (I say nothing of the Sun to find the center of which man has the tools of the eagle's eye, which is said to be the only animal that can look at it), in the other planets I believe that it is not possible" (p. 11rv).

[46] [Lorenz. Chap. 6] "As we mentioned above, the demonstration of mathematicians is lacking. And how could they with their sight pass through the center of a planet? But to take the center from so far away is impossible" (p. 11r).

[47] A Paduan peasant would have been more likely to refer to the campanile at Venice. Allusion to Cremona is certainly intended to suggest Cremonini.

Natale. This way the whole thing's a ball. Oh, him finding a clever speculation to keep the Mathematicians on their toes.

Matteo. But how do you know it's not something thought of before? Son of a bitch if Witelo didn't know it, way back, from what I've heard told to my Master more than once.[48]

Natale. Shall we go on?

Matteo. Sure.

Natale. Brother, you would have wet yourself laughing at the haggling that went on about A, B, N, O, and God knows what.[49] [Lorenz. Chap. 6]. He meant that parallax is great stuff, but that Mathematicians can't make it work.

Matteo. He must not even understand what he was talking about. Come a little closer to me. See that little willow, over by the creek?

Natale. I do, yes.

Matteo. Then you see the poplar, close to the bank?

Natale. Which, the big one? Or the little?

Matteo. The little poplar.

Natale. Yes, of course I see it.

Matteo. Good. Now, look straight ahead. Which of the two looks to you as on the right, the willow or the poplar?

Natale. From here I'd say the poplar is on the right.

Matteo. Now come over this way.

Natale. I'm coming.

Matteo. Stop here. Now what?

Natale. Holy smoke! This way the willow would be to the right and the poplar on the left.

Matteo. Now then, does it matter to you if you can't see through the willow, or the poplar either? You can't see what is right behind them, but what harm does that do?

Natale. None at all, since I look along the edge of the bark; I don't look along what I can't see.

Matteo. And that's the way we look through the air above. Well, that's one kind of parallax. Now come back here.

Natale. Here I am.

Matteo. Look through the top of the willow. Can you see the top of the poplar that I tell you is right in the middle?

Natale. Let me look. No, it may be there, but I don't see it.

Matteo. And if you were so far back that looking over the top of the willow you thought you were looking right at the middle, and you did not see the poplar by raising your eyes, then which of the two would you say was the higher?

---

[48] [Witelo bk. 1 prop. 18, bk. 4 props 51, 66, 67, 70; Euclid Optics props. 23, 24, 28].

[49] Lorenzini's argument was in fact difficult to follow because of the lack of a diagram which seems to have been omitted by the printer in both editions of the *Discourse*. In Capra's reply a diagram was supplied; cf. *Opere* II, p. 295.

Natale. Wait till I think a minute... No doubt I'd say the poplar was shorter and the willow taller, because it would look that way to me, even though it isn't true.

Matteo. Now try something else for a minute. Climb high up on this walnut tree. Here, I'll give you a hand.

Natale. What do you want me to do?

Matteo. Go on up, then I'll tell you.

Natale. All right, I'll go, since you want me to.

Matteo. Be careful though, don't hurt yourself.

Natale. Oh, hell. I nearly lost a fingernail and I've skinned my knee.

Matteo. Well, are you all set?

Natale. Yes, I am; now what?

Matteo. Look again at the poplar you saw from down here.

Natale. Then what?

Matteo. Looking straight at that, can you still see the willow like you did from down below?

Natale. No, but if I was farther off and this high up I'd say the willow was lower than I was.[50]

Matteo. All right. Come on down and I'll tell you a few things.

Natale. It's no trouble getting down (Fig. 3.1).

Matteo. Good; now listen. When you were down here near the willow this willow looked to you taller than that poplar. From up in the walnut tree above the willow things looked just the opposite to you, because this is parallax too, another kind. Parallax just means a difference of viewpoint. Now think. If you went as high up that mulberry over there across the creek, the willow would look lower than the poplar and to the right of it, though back over here at the foot of the walnut the willow looks higher than the poplar and left of it. This is also a kind of parallax as my Master taught me. Now do you catch on?

Natale. I got it, and it's more visible than a barn. I wonder why that fellow in the book only talked of one kind of parallax, when there are three.[51]

Matteo. It's too much for him to think straight. Well now, start figuring. If the new star and the Moon were real close to us, like that willow, the stars higher up would be a lot farther away proportionally than that poplar is. Then would it be possible that no difference of parallax showed up among Spaniards and Germans and Neapolitans?

---

[50] The willow tree was below Natale's line of sight and was hidden from him by branches of the walnut tree beneath him. Had he been equally high but at a greater distance from the willow, that would not have been the case.

[51] There is a reason for this seemingly unnecessarily detailed discussion of possible parallactic effects in a simple rural dialogue. At the time Galileo was writing it, he expected soon to be able to announce and explain in a serious treatise the parallactic displacement of the nova which seemed bound to occur in time. This elaborate discussion of parallax prepared the l reader for his explanation when the event determined whether or not heliocentric motions existed. Displacement due to motion of the observer about the Sun would be analogous to Natale's first move, from N_1 to N_2 in the diagram here supplied. Displacement of the star away from the observer by its own motion would be analogous to Natale's second movement, from N_3 to N_3'. Finally, since the star did not lie in the plane of the ecliptic, its motion would be at an angle northward, analogous to Natale's ascent of the real and the imagined walnut trees at N_4 and N_4'.

RIVER

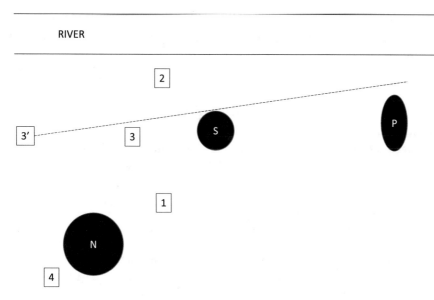

Fig. 3.1 Scheme of Natale's observations (the figure is not present in Galilei's text). The willow is identified by the letter S, the poplar by P, the walnut by N. Natale's observation points are, in order, 1, 2, 3 and 4; of 3' only the possibility is discussed

Yet they all saw it in the same place, near those stars they call the Sagittarius, or the Archer with Crossbow, or something.[52]

Natale. Your argument breaks down, though, because it is not possible to know the distance to the Moon. That fellow says this, too, in his booklet.[53]

Matteo. Not possible? Not his way, but Mathematicians know well enough how to find that.

Natale. I don't know what to say to you, except that you've got reasons to spare.

Matteo. But do you think that fellow with his book would say so, too?

Natale. He'd better, but it may be that such hardheaded to stand on his opinion.

Matteo. Let him settle up, then, and credit my account.

Natale. Next, though I don't know how he could possibly say it, there's something else you're about to hear. Or didn't he say that only plumb overhead could the Sun play hide-and seek?[54] [Lorenz. Chap. 6] But I've seen it [eclipsed], and I know I have.

---

[52] About 17.5° in Sagittarius and 1.6° north from the ecliptic.

[53] [Lorenz. Chap. 5] "Almost contiguous to the lunar orb, whose distance is impossible to comprehend, since this could not be known otherwise than by measuring the shadow of the Earth in the eclipse of the Moon, since the size of the Earth is already known; but to know the size of that shadow is impossible for humans, because it would require to know the size of the Sun and its distance from the Earth, things that cannot be reached with the intellect" (p. 10rv).

[54] That is, be in total eclipse. This is in reply to a particularly silly argument of Cremonini's in Chap. 6 of Lorenzini's Discourse.

Matteo. If the crows haven't plucked out his eyes first, he can explain that next October, since from what I hear, that is going to happen then. But what reason did he give for saying this?

Natale. The Moon wanders a lot, he said, and is only seen straight on when at the zenith.

Matteo. Well? Go on.

Natale. And he says that unless the Moon is at the zenith she can't hide all of the Sun [Lorenz. Chap. 6].

Matteo. The hell you say! This stupid man must think the Moon is flat like an omelette. What an ass! How could more of her be seen by us, she being round [like a ball], by her being at the zenith?—Is there more?

Natale. Sure. What's the meaning of "graxaly"?

Matteo. How's that? Graxaly?

Natale. He says it's cloudy, like milk, and down near the Moon, not really up in the sky.[55]

Matteo. Oh, now I know. [Galaxy.] It's the Road to Rome [Milky Way].[56]

Natale. Oh, sure. Yes, the Road to Rome.

Matteo. That's it. And he says that isn't in the sky?

Natale. No, he says not.

Matteo. Oh, cancer! Why would we call it the Road to Rome, meaning the road to Paradise, if it's not really up there?

Natale. You tell me. Next he takes on something fierce against a Philosopher (even one of the old ones) who didn't believe that it was in the Air because Aristotle told him it was there.[57]

Matteo. Well, let's start home now that it's evening. Anyway we can go on talking while we walk.

---

[55] [Lorenz. Chap. 7] "Being very far from the Poles and the zodiac, it looks like a band, and all the more so because in this region there are many bright stars that with their heat pull it to itself, and make it similar to the stars as a sparkle. It is fixed in its site and as the fixed stars has a movement every 24 hours from east to west" (p. 3v). "Receiving the light of the stars, because of the mediocre refraction, it produces whiteness, like milk or like absorbent cotton" (p. 21v); "and since the Galaxy is near, it will easily be able to draw nourishment from it, so to speak, the more so since it is in Sagittarius near the source of that milk" (Chap. XI, p. 29r). [Editor's note: the constellation Sagittarius is in the center of the Milky Way, where most of the stars are located].

[56] Natale said Grassalia for galassia, galaxy, the word he had heard from the Lorenzini book meaning the Milky Way, called by peasants "the Road to Rome".

[57] [Lorenz. Chap. 5] "Therefore, O Averroes, not being able, or rather not knowing, how to solve this difficulty, you dared to assert that Master Aristotle was no different from mathematicians, and then you said: 'I have seen the stars that are in the Milky Way from different regions of the Earth and they seem to have the same position, and I have observed the star of Aquila, which is at the extremity of the Galaxy, from Cordova and Morocco; these places are very far apart, and I found that the said star was in the same position with respect to the Galaxy.' But the Master spoke openly, and said that the Milky Way is in the air, and no Averro'e can spoil natural philosophy by admitting generation out of the lower heavens" (pp. 12v–13r).

Natale. Go ahead, I'm coming.—Hey, there's something else. He says[58] the new star twinkles because it sputters as it goes around.

Matteo. What do you think about that?

Natale. I'd believe it, except that a lot of stars go around that do not twinkle for me. Instead, I think only the ones that are very high up twinkle, because we can't fix on them well. So this one, twinkling, must be way up there.

Matteo. Move along; what are you, some kind of a rear guard?

Natale. And so, since that fellow doesn't know where this star is, and can't know how it was created, everything he says about it all comes out just so much blather. Is that so?

Matteo. It must be.

Natale. All right, let's have a little fun with the forecasts of his, that's what I'd like.

Matteo. Sure. What does he say?

Natale. He says the star will last quite a while, unless broken up by the Sun; that's what he says.[59]

Matteo. He might as well say just that it will last till it breaks up. Anyway, when it is gone, he can say that he broke it up.

Natale. Plague take him, that would be pigheaded. Then he says there[60] will be plenty of everything, and that this is one of those lucky stars.

Matteo. For those who have things going well for them. But who will that brief this drought keeps up? Still, believe him in your own way.

Natale. About humankind, then? That may be.

Matteo. What would that mean?

Natale. It would mean that humans will become clever and wise, and will stick to the truth [Lorenz. Chap. 12].[61]

---

[58] [Lorenz. Chap. 10] "Then we must say of its scintillation, or tremor, which in the fixed spheres occurs because of distance [...]. But I will add another reason that applies to the abovementioned star, and it is due to the matter of which it is composed. Because of its very rapid movement it becomes a fan, and the air generates that subtle glow of fire, as when we see the flame awakened by the bellows" (p. 27rv).

[59] [Lorenz. Chap. 11] "We often talk about predictions, and first of all about the new star. Some are of the opinion that it will not last long and, having come so quickly to such greatness, it will fade away in a short time; then, when the rays of the Sun will reach it, it will dissolve because of the heat of the Sun accompanied by that of Mars and Jupiter" (pp. 28v–29r).

[60] [Lorenz. Chap. 11] "But this star has borne the good qualities of a simple, temperate, spiritual, and penetrative substance, endowed with efficacious virtue, and pure white, shining softly, mild, well gathered, and lofty, fallen in every place on Earth, infusing on all a sweet subtle warmth. These stars, this one of today as well as the one of 1572 in Cassiopeia, communicate little by little their nature to our inferior one with their benign breath; they fight humidity and dryness with their sweet water and their comfort; so they make fertile and healthy the Earth, and healthy the animals. Therefore positive effects are to be expected" (p. 30v).

[61] "Anyone could say that the above-mentioned star is primarily for the purpose of purifying the bodies, senses, and intellect of men, so that the wisdom and contemplation are of great profit to drive away bad opinions, and every kind of ignorance, especially the malicious one; and that and the illustrious arts already fallen will rise again, because the dryness, and the subtlety of the humors gives much to the spirits and the intellect" (p. 31). "I will also say, however, that the new star of 1572

Matteo. Just see how that forecast worked out in his case. You see how well that is, don't you? There's no getting away from it, his having had such a brain from childhood.[62]

Natale. You're pulling my leg, aren't you? Instead, the forecast applies to us because we stick to the truth even though he tried to swindle us.

Matteo. Hurry up, you aren't winded.

Natale. Next, he says that the star takes away trickery and madness, but I don't know about that.[63]

Matteo. Yes, of course, our affairs will be none the worse for that. But his forecasts don't surprise me, since his whole book seems to me like so much fortune-telling, always divining things.

Natale. Here he says he has another book, in Latin,[64] ready to print.[65]

Matteo. He should get on with it, because it's nearly Lent. It should be good for something, because this one that came out during Carnival has given us a lot of laugh.

Natale. That man who was reading said he thought the other one would really be printed to sell copies and make money.

Matteo. May he work to hurry it along, then. And if he has extras of this one, let him make a bundle and throw it where Tofano[66] threw the spices, to be put to good use.

Natale. Here we are, home; let's lay off. Will you have supper with me? You know I'd like to have you.

Matteo. I know. But I can't, because Menica[67] is waiting for me. But thanks anyhow.

Natale. So long, then.

Matteo. Goodbye.

THE END.

---

in Cassiopeia, born beautiful and high as well, was followed shortly afterwards in some country of Italy and in other countries by pestilence. But perhaps this happened by chance" (pp. 33v–34r).

[62] Cremonini had been a child prodigy.

[63] [Lorenz. Chap. 12] "So that wisdom and contemplation will be of great profit in driving out evil opinions, and all manner of ignorance, especially malicious ignorance" (p. 31).

[64] Antonio Lorenzini, *De numero. ordine, et motu coelorum* (Paris, 1605).

[65] [Lorenz. Chaps. 5, 6] "As discussed in my book on heavenly things against mathematicians, in Latin, which will be published shortly" (p. 4).

[66] Tofano was a stock figure in the popular theater. This name appeared in one of Galileo's draft plays (*Opere* IX, p. 200) as that of a poor merchant. The allusion here seems to be an outhouse; had a poor merchant needed to hide smuggled spices from the tax collector, he would have put the package into a smelly place.

[67] Menica (for Domenica) was also a stock character, and the name likewise appeared in a draft play by Galileo (*Opere* IX, pp. 198–9.)

Stanzas by an Unknown Author to the New Star Against Aristotle
What raving, o foolish Stagirite,
your belief that creations are forbidden in Heaven!
A newborn star, in it fixed, its clear face
shows glittering, and you don't see it?
Or you deny sense more than ever
getting stuck in your first mistakes, and ask for other proofs?
But how to deny this simple evidence
when all your science was based on sense?

That in Heaven the star was born
maybe you think not to be true;
though every nation sees it
fixed in the same point up on the sky.
Everybody sees it in the beautiful location
where Sagittarius has its splendor separated
and as if from Scorpion's tail escaping
rises trembling, and it is nailed over it.

Now if it be under the Heavens, under the Moon,
so far from the stars of the sublime roof,
as of another climate, how it does not partly hide,
or at least changes its aspect,
and in a variety of sites seems now
in the same constellation to have shelter?
Indeed, how does it return to us in the morning
where it appeared in the evening?

Nor here cease my true words,
which firm and new now return.
Tell me, if the Earth
turns twenty-two thousand miles around
and this new light, like for the Sun
turns it all round in a night and a day,
how could it for three hours and more
on our horizon show its face at sunset?

It cannot be that, if it is so humble,
it sets its pace as to be seen,
since in a short time
it would become invisible;

nor it could look always similar, unless, as it lengthens from us,
it does acquire mass
and then loses it while it draws near
to look the same in bulk at every place.

But since it is fixed in the sublime sky,
Its height proves so firmly, that
as placed between the first stars
naturally shows itself there and moves accordingly;
nor it could be found in lower parts
varying in size or site,
since its place on Earth does not change
given the enormous circumference of the Heavens.

Therefore, ripping the cloth blinding you,
now let your tongue melt into true notes:
if you believed, at your scorn and harm
the upper wheels to be immutable,
now that this new flame shows your mistake,
you know that Heaven can be generated,
and thank Mother Nature,
who sent you such graceful light of truth.

The Controversy with Baldassarre Capra

On February 16, 1605 was published in Padua at the publisher Pasquati the *Consideratione astronomica sopra la nova et portentosa stella che nell'anno 1604 a dì 10 ottobre apparse, on un breve giudicio delli suoi significati (Astronomical consideration on the new star appeared on October 10, 1604, with a short judgment of its meanings,* written by the astronomer and mathematician Baldassarre Capra (who defined himself on the cover page as "astronomer and medical doctor"). Baldassarre (Baldesar) Capra (1580–1626) claimed the priority of the discovery of the supernova in controversy with Galilei.

About the position of the star, Capra had no doubts: in no way it could be under the Moon. But, while repeating what Galilei had said during his public lectures (which Capra had attended) in explaining how the absence of parallax was indicative of an enormous distance, Capra attributed to the star an extremely small parallax, so small that it could not be detected (especially with Galilei's techniques), but not zero.

Capra wondered what could have originated the star, and admitted that he was unable to explain it. He was aware that the Aristotelian hypotheses were contradicted by the event and that there was a need to find other explanations, but he confessed that he was not able to find any. He also speculated on the meaning of the star, in particular in relation to the previous new star that had appeared in 1572 in the constellation of Cassiopeia. He observed that the two stars had appeared exactly 32 years apart, a number of years equal to the age of Christ at the crucifixion; this led him to consider that both stars were bearers of great and wonderful events, "like the comet seen by the Wise Kings in conjunction with the birth of Christ, which foretold the destruction of Judea and the conversion to the true faith".

Capra addressed three accusations to Galileo, thus starting a controversy.

1. He accused Galilei of being inaccurate about the exact date of the apparition of the star. In his lectures, Galilei had reported that the star had appeared between 8 and 10 October,[68] while Capra was sure that the correct date was the 10th.
2. Capra interpreted that inaccuracy on the date as a clear intention on Galilei's part to take away his recognition for having seen first in all Padua the star nova. Capra knew that Galilei had not observed the supernova until October 28, and therefore accused him of claiming merits that could not be his.
3. He stated that Galilei, in his first day of observation, had completely misunderstood the position where the supernova was located. According to Capra, Galilei had placed the new star on the straight line joining the tail of the Swan to the Boreal Crown, while at that moment the star was on the straight line joining Mars and Jupiter.

At first Galilei did not respond publicly, but limited himself to annotating his personal copy of Capra's pamphlet with harsh and often vulgar comments.

We report here a translation of Capra's pamphlet, with Galilei's annotations (indicated by Roman numerals).

---

[68] In his preparatory notes Galilei wrote "October 10", but during the public exhibition he did not exclude the possibility that the new star had already been seen in the nights immediately preceding October 10. The question of the date was resolved after an exchange of letters between Galilei and Altobelli: it was October 9.

Capra's *Consideratione astronomica circa la nova, e portentosa stella...* annotated by Galilei

To the Most Illustrious Uncle
and Observant Master
SIGNOR GIOVANNI ANTONIO DALLA CROCE.

Not differently from you, who in the desert, raising your eyes to heaven, complained that you remained the only worshipper of the true God and preserver of the true religion, I, in the desert of my voluntary exile from my homeland, have often complained, believing myself to be the only lover and defender of the[69] mathematical sciences against the ignorant slanderers. But in the end, just as Elias was told by the Divine Goodness that he should know that he was not the only true believer, since God had reserved for him seven thousand men who had not been contaminated by the idolatry of the Idol of Baal, so I have been persuaded that I am not the only protector of mathematics and the sciences, especially since I remembered that I had You as my Uncle[70]: You, a person in whom all virtues and the desire that the virtuous be exalted respond to the highest degree. I was therefore consoled and I was bolder in committing myself to speak of such a portentous marvel,[71] and in part even in rejecting what had been said about by mathematicians: so that, by consecrating to You this small fruit of my studies, You might come to know me as a nephew[72] and faithful servant. I ask You to receive it as kindly as from the hands of one who loves You cordially, and who in so doing will find the courage to do even better things. I will not omit to say that if one considers the person to whom this little work is dedicated, or if one considers the subject it deals with, it should for every reason have been written in Latin as more excellent and worthy; but since those who have opposed mathematics have written in our mother tongue perhaps in order to put this science under suspicion among the ignorants, although such doubt could not have fallen on the learned, you will excuse me[73] if, wishing to make known to all the objections I have made, I have written again in the vernacular. Since I recognize in You that humanity and sincerity which have shone in all Your previous writings, I will excuse myself.

I will not be long in offering you these labors of mine and in excusing myself for some of my imperfections. I humbly kiss Your hands and ask Our Lord for your greatest good.

In Padua, on 16 February 1605.
Your nephew and most affectionate servant
Baldassarre Capra.

---

[69] Here Galilei deletes âŁœlover and defender ofâŁž and writes in place of the deleted words âŁœjust jerk in theâŁž[note of the Editor: the original word for jerk is coglione, i.e., âŁˆtesticleâŁ™].

[70] Galilei's note: "Oh great man this must have been, who could not remember having an uncle."

[71] Galilei's note: "And what portent, since you never said of what you want to talk?"

[72] Galilei's note: "Nice parenthood: the uncle does not know the nephew, nor the nephew the uncle. Beasts."

[73] Galilei's note: "I have excused you too much. Please do not get tired, for I see that you don't know how to speak in the vernacular, nor in Latin.

As I wondered if I should write something about this portentous and no longer observed star, which appeared in October of the year 1604, many reasons persuaded me to do it, because having sustained so many labors, vigils, and discomforts (of the body as well as of the mind) in order to observe it diligently and to know its place and its nature, and having faced not a few expenses in building instruments for such an operation, it seemed to me a convenient thing to show to my friends, and to others who were aware of my efforts, that they had not been thrown to the wind, but had brought me contentment and results. But on the other hand I was horrified to see such a sharp contradiction between natural philosophers and mathematicians. The latter say that most comets and all stars are generated in the firmament, while the latter deny any alteration of the sky and persist in their opinion that they are generated in the region of the elements,[74] believing that they would do Aristotle a great wrong if they admitted something against his opinion, since it is no more opportune for a natural philosopher to investigate the causes of things than to defend the opinion of his master, even in a matter already confirmed twice[75] and diligently observed. While I was considering all this to myself, having seen that the most excellent Galileo, in his learned lectures that he gave publicly on this star in the past days, did not want to speak openly about the time of the appearance of this star, nor about its location in the zodiac, but confusedly said that it was located in about 18° of Sagittarius with almost two degrees of boreal latitude, I took courage, hoping that in the future he will be able to give a more accurate description.

That little bit of courage became a very ardent desire that convinced me to realize as soon as possible what I had decided to do after I saw a *Discourse* published[76] about this new star, in which besides the fact that at the beginning it does not tell the whole truth about its appearance, in the following it induces us to wonder, engineering with new theorems to reject the validity of the parallax measurements made diligently by mathematicians. Having therefore decided to write my opinion, I proposed to examine in part this Discourse persuading myself that I had a good opportunity to demonstrate what I had in mind to propose on this star: which I decided to do not out of a spirit of contradiction, which has always been alien to me, but out of pure zeal to know the truth, which only by doubting can be discovered. I hope that the author will not be offended that I examine his new theorems; on the contrary, he should examine these writings of mine, and if he finds something worthy of correction, kindly inform me, because I will always be ready to change my opinion.

I will first consider the time of the apparition proposed by the author with the other circumstances, and then determine what was the true time of the first apparition; I will then go on to consider what he says against the parallaxes, even if, with the author's good grace, it seems to me nonsense; I will touch again on some points that seem to me noteworthy. Therefore thanks to the diligence of my observations, both

---

[74] The lower spheres in which the elements water, air, earth and fire reside.

[75] It refers to the supernova of 1572 and the Great Comet of 1577. This last is a non-periodic comet that was well visible to all of Europe, and was studied by Tycho Brahe and by the Turkish astronomer Taqi ad-Din. Brahe located it above the Earth's atmosphere.

[76] He is talking about the *Discorso* di Lorenzini.

when this star was in the West and now that it is in the East, I will clearly indicate its position with respect to the ecliptic and to the Universe. I will conclude by saying something about the effects the star can have.

The author of the *Discourse...* proposes that this star has been observed in October of the year 1604, on the 8th approximately, in the 18th degree of Sagittarius. I would like to understand what this "approximately" refers to,[77] because it can refer to day 8, as well as to 18°. Attributing it to day 8 would be a too indefinite proposition, because it could also say that it appeared in a day of October; if we attribute it instead to 18° it seems to me also a great generality: but perhaps he will answer me, that not being an astrologer, he has not observed, and therefore he cannot know the true day of the first apparition, and a more precise place than that which has been published by those who have observed; which is granted to him gladly, since as is evident from his writings he does not occupy himself too much with mathematical things. Instead we cannot accept this lack in the most excellent Galileo, who in his lectures confused the eighth day with the ninth and tenth, so that it was not possible to understand whether this star appeared on the eighth, ninth or tenth day, a fact that should have been diligently described; just as he also proposed without any precision the place with respect to the ecliptic. But coming now to the determination of the day on which it appeared for the first time, I will tell you that according to my habit (which was to observe every day[78] both wandering and fixed stars), I met with the German Simon Marius, my dear teacher in this activity, and with the Calabrian gentleman Camillo Sasso, on the tenth day of October, to observe Mars, Jupiter and Saturn. While I was preparing a quadrant to measure the heights of some fixed stars, Simon was looking in the direction of the new star, which almost made a perpendicular line with Mars and Jupiter. Camillo, although inexperienced of astrological things, raised his voice saying that there was a star he had never seen. Simone came towards me shouting, "We have a new star!". I reached them and observed a star of color and size similar in every way to Mars, which was not there before, a fact of which I was sure, having on the eighth day and the preceding ones always observed the said planets, and particularly having on the third of October observed a star of fourth magnitude, which was only 49 min distant from Mars; so that I immediately conjectured that this star had appeared between the ninth day of October and the tenth. But on October 9th the weather was cloudy, so the stars could not be seen, and therefore whoever says that this new star was discovered here in Padova before the tenth day is wrong. So, after having seen this star, and in the same evening having also observed it by measuring its distances from some fixed stars, for the following five days it was impossible to see it because of the intermittent and continuous rain. On October 15, finally, the sky became clear again, and it was possible to see the aforesaid star, which appeared to be of higher magnitude, like that of Jupiter and perhaps a little more. Its color, although similar

---

[77] Galilei's note: "My ox, I will tell you. That approximately refers to days; but what do you mean by that? Do you not see that this is said modestly? For who would assert that that the new star was observed at such a moment in time, on such a day, and could have been seen even a little earlier by others not known to the author? Etc.

[78] Galilei's note: "You, who are pedantic about the word approximately, should have said every night, because at night the stars are observed."

to that of Mars, also had much of the Jovian color, and shone above every fixed star. From this it may be deduced that it is not entirely true what is said, namely that this star has increased in magnitude from day to day; for if on the fifteenth day it appeared greater, but it didn't increase any more. If approaching the Sun appeared a little smaller, it was not because its magnitude was diminished, but because the light of the Sun obscured it as it happens for all other stars: moreover it turns out to be false that this star was similar in magnitude to Venus, not exceeding Jupiter except by a little.[79] Therefore, after the aforesaid miracle was observed again on the 15th, going one day to visit the illustrious Iacopo Alvise Cornaro, I informed him of this new and peregrine light; he showed a great desire to see it, and the following day, I believe, finding me again in his house, he asked me the position of this star with respect to Mars and Jupiter, asserting that he wanted Galileo to see it; I, who up to that moment had no precise measure of the position of this star, reported it as about 18° in Sagittarius, and its latitude as about two degrees to the boreal side, and I also described to him the position with respect to Jupiter and Mars. After a short time I was told by Cornaro that the most excellent Galileo had then seen the peregrine star. From this we can draw a necessary conclusion, namely that Galileo had known the time and the place to observe this new wonder thanks to Cornaro, but he did not mention it in his lectures. I believe that the above story clearly shows that this star was not observed by anyone before the tenth day, since in the ninth day it was not possible for anyone to see it, even if they wanted to believe that it had already been generated at that time.

Coming now to the part in which our author speaks about parallaxes, it should be noted that in this chapter, in his desire to describe the opinion of mathematicians, he speaks generically about parallax, but not like mathematicians, who subtly consider parallax according to the length and width with respect to the ecliptic, but this can be granted to him, since as a philosopher he knows little about such things. Leaving aside parallax in Chap. 4, it is stated that mathematicians contradict themselves, because they cannot answer the question of how generation takes place in the sky, how this star was generated, why in such a long time no part of the sky has ever been corrupted, nor other similar questions. It is enough for now that mathematicians prove that this star is in the starry sky, as it was already proved by the most noble, learned and brilliant Tycho Brahe for the star of the year 1572. It is then up to you natural philosophers to solve the proposed questions and to find the mechanism of these celestial generations, as undoubtedly would do, if he were now alive, that very gifted and brilliant philosopher Aristotle, who diligently considered everything that had been observed by the mathematicians in his epoch.

---

[79] Galilei's note: "And why should I believe you, who make it little greater than Jupiter, more than those who compare it in size to Venus?"

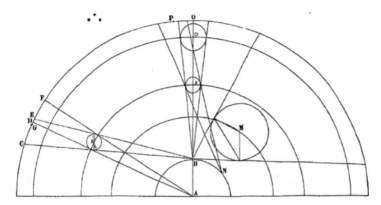

Beginning in the first Chap. 6 to speak at length against parallax, the author of the *Discourse...* clearly shows that he has no knowledge of how mathematicians measure the parallax of celestial bodies; and I marvel at this kind of argument when he says that the visual ray does not pass through the center of the planet, and therefore parallax cannot be observed. If this author had been educated in the mathematical school, he would have seen that parallax makes all a celestial body appear in another place, and not only its center; he would have also learned that it is not so difficult to find the center of any celestial body, even a small one. As for the example of the Moon, when he says that it would not be surprising if the holes intersecting the center of the Moon in various parts of it formed various angles (as if he wanted to say that the parallax observed by mathematicians is nothing but that difference that in this figure is indicated by GD, which is the difference between EF and CD, which is not the true parallax observed by mathematicians, but the difference between two parallaxes). In the figure point A is the center of the Earth, point B the eye observing the Moon, EF the parallax observed in the upper part of the Moon's body, CD the parallax observed in the lower part of the Moon. If parallax is observed in the upper and lower part of the Moon, I do not see why the center should not have its parallax. It follows that the difference in appearance of the Moon cannot be explained by the variation of parallax angles, and less for the other celestial bodies, which do not appear so large.

Who says that the center of the Sun is impossible to find because of its splendor clearly shows that he has not practiced much astronomy and mathematics: certainly if he had knowledge of them he would not believe mathematicians so much. Fools who, wanting to find the center of the Sun, would try to sharpen their eyesight by means of its rays: wanting to learn from mathematicians, they would instead find it very easy to investigate the true center of the Sun. He then goes on to say that it is impossible to know whether Venus hides Mercury, or Mercury Venus, and so of any other star, and so he would make sure that instead it is well possible, as did that very diligent mathematician and observer Simon Marius, who had observed that the Moon eclipsed Mars for an hour without interruption.

It would take a lot of time to examine in detail the demonstration in the second Chap. 6, but knowing that the truth is satisfied with a few words, I will strive, as briefly as possible, to grasp its meaning, and to see if it is of such value as to deserve the sacrifice of a hundred calves,[80] or if, as I believe, is based on weak principles, not to say false. I would be glad to know why he has not proposed a figure to illustrate this demonstration of his: if he has not done so believing that it would be easy to understand it, it is worthy of apology; but if he has done it in such a way that it could not be fully understood, it would not seem to me a good thing. He therefore proposes as the basis of his demonstration that two lines coming out of an eye cannot touch a spherical body except when it is at the zenith, because only there those lines can form a right angle with the lines drawn from the center of that round body. This is false: a spherical body placed in any other place than the zenith can be touched by two lines coming out of one eye. I do not think that this needs a demonstration. If I say that this is obvious, it is now useful in the above figure to show openly that not only when the spherical body is at the zenith, but also in any other place, it can be touched by two lines coming out of the same eye, since they can make two right angles with the lines drawn from the center of the circle, as shown by Euclid, lib. 8, prop. 18 and 19. Let us therefore place the spherical body at H and the eye at the point B from which two lines BC and BE are drawn; I say that these two lines, forming right angles with the lines I and L drawn from the center, will necessarily touch this spherical body H even if it is placed very far from the zenith, which is true not only when one sees the whole half of the body, but also when one sees only one of its portions, as appears in the figure indicated by the letter M. From which it follows that the foundation of his demonstration is false. Moreover, we deduce as manifest falsehood that the eye placed at point N cannot send two lines touching the Moon if it is not at the same height of the eye placed at point B, because wherever you want to place it, it can always form two right angles as it has been said above.

Having therefore destroyed this foundation of necessity, the first corollary falls, namely that the perfect solar eclipse cannot be observed except by those who live under the ecliptic. But I do not know what to say, seeing that so many words are used to destroy the parallax. Moreover, it seems to me surprising that he says that when the Moon is seen both from the point B and from the point N under the Sun there is no parallax; therefore it seems that the author does not understand what parallax is, nor how mathematicians measure the position of planets or other celestial bodies: it is true that everyone of those who are in B as well as those who are in N see the Moon under the Sun, but this does not mean that the Moon is under the Sun. is the place considered by mathematicians when they talk about parallax, because those who are in B consider the position of the Moon in the same point of the extreme sphere with the Sun, in O, but those who are in N consider the place of the Moon in the eighth sphere a little further away, that is in P, and this is the reason why in various circles the same eclipse varies in magnitude. But since these things are so trivial that they will move to laughter the mathematicians who will read them, it will not be out of purpose to move on to something else, if first I noticed something frivolous. Our author says, to confirm his reasons, that the crowns that sometimes

---

[80] According to a legend, Pythagoras celebrated the demonstration of his theorem by sacrificing to the gods one hundred oxen.

shine around some stars are undoubtedly placed in the air, which is true, but they are seen everywhere in the same way. If he had well considered what Aristotle adduces as the cause of these crowns, he would certainly not have given such a weak reason. In fact these crowns are nothing else than a refraction of the light of the stars in the humidity of the air, and therefore it happens that they are seen in every region, provided that in every place the air is disposed to receive this refraction, so that if it happens, as we often see, that one place is disposed to receive this refraction and another place it is not, then this corona is seen in one place and not in another and even when it is seen in each place, it is not the same corona, but related to different portions of the air.

On the contrary, this new star can be observed from any region even if it is very far away, and I know from letters of an excellent person that have been diligently read by me that in Germany this star has been observed with the same magnitude, with the same colors, and in the same place: from which it can be concluded that the observations of the mathematicians are not so doubtful, since those made in Germany are exactly in agreement with those made here in Padua.

From the aforesaid principles at the end of the chapter the author draws an unjustified consequence, namely that this star cannot be observed by others than those with respect to whom it is vertical, and yet he wants everyone to see it in the same place; whether these consequences deserve reproof, let those who understand them say so.

Having established that this star is located in the region of the elements, and seeing that it is very difficult to persuade oneself that this hot, dry exhalation can be found under the Moon, where according to the Peripatetics' opinion the sphere of fire is located, the author strives in Chap. 7 to prove that the elements are mixed, and therefore declares that there is no simple fire in the star, but fire mixed with air. If this were true, since the matter there is very rare and disintegrated, the other reacting elements would not be able to resist for long: if therefore there is fire there, I do not know how a hot and dry exhalation, made of thin and tenuous parts, can last so long. I find it even more difficult to believe when he says that the hot and dry exhalation that Aristotle attributes to the galaxy does not go under the zodiac because there it is dissipated, nor under the poles. I do not see why this exhalation is not sooner attracted by the stars of the zodiac because they are bigger and more abundant. Moreover, if the heat of the zodiac dissipates the exhalation, I do not know why this new star, which is not only in the zodiac but is even closer to the ecliptic, has not been dissipated so far. As for what he says about the change in the galaxy, it seems to me that it is not easy to conclude against many excellent mathematicians, who have observed and in their writings have left a trace, that the said galaxy has always occupied the same fixed stars (especially drawing this from the writings of those who perhaps have not observed much such things). On the contrary, it would seem to me that the new star, moving with the fixed stars, should certainly indicate something different from an exhalation; which is also confirmed by the fact that when the Moon or another planet passes by it, nothing is lost of its brightness, which would be impossible to believe if it were really an exhalation. But since these arguments of the author are unreasonable, I do not want to add anything.

Turning to the mode of generation of this star, I do not see then why such new stars do not form more often, since the fixed stars always have the same position with respect to each other, and consequently the same union of lights should occur often. But if you related it to the union of the lights of some planets, this is even more expressly false, since it is not possible for the new star to be so immobile, having its cause in the planets which are mobile, especially if it is in the sky of the Moon which, as Aristotle established, would undoubtedly kidnap it. Finally, I do not see why, such congregations of celestial bodies happening every year, they should not produce every year such new stars.

Moreover, in explaining why this star sparkles, among other things he mentions the rapid motion of the sky that ventilates that fire, as our fire is usually moved by the bellows: if this is true, I do not know why the other planets, which are more distant and equally fast, indeed faster, do not make those flashes of light. To this we must add that since this star is now manifestly diminished in size, it should of necessity have partly lost its sparkle; instead, sparkling as before, it manifests the fact that it is not a fiery exhalation. Explaining then those spots which appear on the Moon, he says that they are nothing but vapors, which partly dissipate with the light of the Moon, and partly remain there; but have you seen these spots in the unilluminated part of the Moon? It is not possible that the almost completely opaque part of the Moon can disintegrate and undo those vapors, which, disintegrated in the shining body of the Moon, according to his opinion, form the aforesaid spots.

Finally, having arrived at the place where I have to make a note of what I have observed myself, I make a remark about that part of Chap. 11 where he speaks of the duration of this star, and I ask how it is possible that this star, being an exhalation, has been under the solar rays for almost a month, and has not been dissipated by that light, if, as he says, the reason why the galaxy is not under the zodiac is because the exhalation there is dissipated by the light of those stars. If the light of those stars can dissipate that exhalation, for what reason the light of the Sun, which in proportion is much greater, could not dissipate this vapor? I do not know if this is a way of philosophizing or of mocking.

This is what I think of this *Discourse,* and I say it not with the intention of opposing it, but only for the love of science, with the little intellect that nature has granted me. It will be up to you, excellent gentlemen, professors and mathematicians, to defend the noble doctrine of mathematics from the hands of those who want to destroy it; I am sure that you will not lack reasons, but the passion I have for this noble science is so great that I am forced to beg you earnestly not to fail, so that certain people will have no reason to persist longer in wrong opinions.

Summarizing therefore my reasoning, as I said above, this star appeared on the tenth day of October. It was in every way similar to Mars both in color and size, and admirably twinkling; so that I was astonished, and I could not stop (I was certain from my observations that such a star had never been observed in that place, and the fact was also guaranteed to me by Simon) diligently examining all the catalogs of fixed stars. In spite of this, to tell the truth, I remained doubtful until the 15th, on which day I did not wish to wait for the setting of the Sun, but endeavored to find out if I could see the new star; I saw it of the size of Jupiter, or a little bigger, and it had somewhat abandoned that rosy glow which it had at first, and shone with a mixed color of Mars and Jupiter. At that moment I left all residual doubt that this star was one of the known ones, and devoted myself to observe it.

And having spoken of the site of this star, many times for half an hour before sunset we watched it in the presence of many friends. On the third of November, in the presence of Paolo Boim, mayor of the University of physicians of Padua, who had already taken part in the observations many other times being particularly fond of such science, this star was seen to twinkle as the rays of the Sun struck our eyes. Finally, as the Sun approached the new star, it appeared a little smaller, which I do not think was in any way because it had diminished in size, but because the greater light of the Sun began to obscure it; and towards the end of November, partly because of the presence of the Sun, partly because of the clouds that were on the horizon at that time, it was totally obscured. It was conjectured by me that this star, on the feast of the birth of our Savior Jesus Christ, would be seen again in the East; so it was observed on the eve of this feast by the most excellent Galileo. When it was raised and moved away from the Sun to such an altitude that it could be observed, it was diligently studied by me up to the present day with instruments free from all error, as it will still be, God permitting, until its consummation. It did not, however, appear with its former magnitude, as can be clearly seen, but is fixed in its proper place where it had been observed while in the West, as I shall later say, and not a little sparkling.

As I promised, I want to show faithfully and precisely, not with words but with facts, the position of this star. I will start by describing how I found the immobility of this star both with respect to what fixed stars and by measuring the longitude and latitude with respect to the ecliptic. When this new star appeared I found only one instrument with which I could measure the distances between the stars; it was not very large, but whether it served faithfully will be judged by the reader from what I am about to say. Moved then by this new excellent spectacle, judging it worthy of a better instrument, with all possible diligence I had a sextant[81] made in the likeness of the instruments of the most noble and excellent mathematician Tycho Brahe. In the meantime, however, I made use of the first instrument until November 6, the date on which the sextant was made. Then I observed the distance between this star and two other fixed ones, that is, the brighter one from the right foot of the serpentarium, and the other brighter from the left foot of the same, and many times having repeated the said distances with all possible diligence, and especially while not only the new star but also the two aforesaid fixed ones were very high, and for this reason did not undergo any or little refraction, finally, having made the evaluation with the technique of spherical triangles, I measured the place of this star at a longitude of 17° and 39 min in Sagittarius, and a latitude of one degree and 51 min towards the northern part. After observing until November 6 with the aforesaid instrument, wishing to be sure of the position of this star and its immobility, I began on the 6th to observe with the sextant, and found the same place by taking the distance from the other,

---

[81] A sextant is an astronomical instrument that measures the angular distance between two visible objects.

more distant stars, since the previous ones due to the proximity of the Sun were beginning to become unobservable. I found the place of the new star at 17° and 38 min of longitude in Sagittarius, and the latitude of one degree and 49 min towards the boreal part. From this it can be seen if this star is motionless, if I have faithfully reported what I have been able to observe so far. Perhaps some may doubt that the observations are not correct because of that small difference, which in the latitude is seen to be two minutes, and in the longitude one minute, but I think this may come either from some refraction that the new star has had, or from the fact that, believing I did not need so much precision, I overlooked a few minutes seconds, which are well visible in my sextant. This is enough to show the place of this star with respect to the ecliptic; from where we see that it has never changed place but it is fixed. In this detail it should be noted that Galileo, in his lessons, wanting to prove that this star was fixed, pointed out that he had observed with one of his instruments that this star always formed a straight line with the brightest star of the boreal crown and with the brightest star in the tail of the Swan, which was not possible if this star had some particular movement. I thought for a long time how he could say that these three stars form a straight line, since they form a triangle: at the end I concluded that it was more likely that he was talking about some other star, which was really in a straight line, or that I had misunderstood his words. But supposing that he was speaking of two stars which were really in a straight line with the new one, it is not quite certain to affirm from what he said that this star was motionless, because although when it was somewhat elevated it could form a straight line with the two supposedly fixed ones, then near the horizon by refraction it could no longer form the said straight line; from which one can discover the uncertainty of these instruments with which one wants to measure straight lines. Nevertheless, even with the mode of his instruments which required more care, I commend his intention to strive to show for the public good the accidents of this star. Coming now to the position of this new star in the Universe, I declare that in no way it can be below the Moon, in the region of the elements, as was very well demonstrated in his learned lectures by the excellent Galileo, who said that this star, not having any parallax, should be placed in the eighth sphere, which was found to be true. And I intend scrupulously to demonstrate that it must be placed in the eighth sphere among the other fixed stars also because of its immobility, scintillation and form, together with other similar characteristics; for if it were in the air, which by its nature is vague and fluctuating, and because of the continuous ascension of the exhalations, it would be impossible for it to have reserved its place without changing it. If it were still in one of the aforesaid spheres, as in the sphere of the Moon or of Venus, it would necessarily have been moved by the motion of that planet; especially if we do not suppose that the stars move by their own and indefinite motion, but are surrounded by their own sphere, as Aristotle wants; and the same must be said if it were placed in any of the other spheres. Moreover, since that scintillation is similar in all respects to that of the fixed stars, and not to the light of the other celestial bodies; since its form is similar in all respects to that of the fixed stars, and not to a burning flame, and it is impossible to believe that it can come from an exhalation, it follows that it cannot be in any other place than among the stars. But above all reasons, from the fact that this star has no parallax, it is evident that it can only be among the fixed stars, for which parallax is not perceptible because of its smallness.[82] This is indicated by the fact that I have always observed the same distances, both when it was close to the horizon and when it was very high up: I have never found a difference greater than five minutes of arc, a difference which came from refraction, indicating that these five minutes made the observation near the horizon shorter than when it was far from it. This is contrary in every way to parallax, which shows that distances near the horizon are greater. This is followed by immobility, scintillation and shape, being under the sun's rays for a month, deprivation of all parallax. This is all I had to say at the moment about the position of this star in the Universe. It will be observed that I have deliberately omitted to demonstrate geometrically that this star has no parallax; since it has not yet reached the meridian, many observations remain to be made, of which, without any doubt, I believe that at the end of this miracle Simon Marius will speak at length. Now those who slandered him for being an astrologer inexperienced of geometry will see if they have told the truth, or if, as is the habit of slanderers, having little or no knowledge, they have tried to damage unfairly the reputation of others.

Having so far shown that this star must necessarily be generated in the sky, it is of great importance that especially natural philosophers should endeavor to discover the mechanism of this generation in the sky, and not to persist in believing that there is no alteration. As far as I am concerned, I do not know how to explain this kind of generation, but I believe that the way proposed by Aristotle, which is appropriate only for the elements, does not fit in any way with these celestial bodies; but that it is necessary to find another way by which these phenomena can be explained. And whoever should find such a way, which I do not think completely impossible, I wish to communicate it to all, because from this will arise eternal glory, not only among philosophers, but also among mathematicians.

With this, then, I seem to have fulfilled my purpose, having demonstrated the true time of the first appearance of this star, and having moreover placed this star in its place with respect to the ecliptic as well as to the Universe. But since those who admire such portents must also be very anxious to know what they mean, I will make some pious considerations, rather than according to the custom of astrologers to compose a decisive judgment.

Therefore, in order to determine more easily the meaning of this star in a conjectural way, and without any superstition, it seems to me convenient to add to this consideration the star that in the year 1572 appeared in the sign of Cassiopeia, because it was similar to this one both in its magnificence and in its position in the eighth sphere, and because it seems to me that these two stars contain in themselves a certain mystery, since the intermediate time between the appearances of those two stars corresponds almost punctually to the age of the our Savior Jesus Christ. I therefore consider that these two stars situated in the highest part of the Universe, that is, in the eighth sphere, may be the herald of some great and wondrous event, and this not in a particular but in a universal sense, which may be confirmed if we consider the immense magnitude they must reasonably have, appearing to us greater than any

---

[82] Galilei's note: "There is no parallax".

other fixed star, with all its great distance. Moreover, I believe that as in the birth of the only Christ appeared that star observed by the Wise Kings, which was the pretext of many ruinous events in Judea, and of the conversion of the Gentiles to the true faith, so perhaps we can say that these two stars are the preludes of some great change in the Universe (even if the stars have no influence in the mysteries of religion, but only meaning). This is confirmed by the fact that the new star of 1572 appeared in the northern part so that the inhabitants of those areas could see that prelude of such a change: but since there were some regions in the southern part of the Earth toward the Antarctic pole that could not see that God-given indication, now another appears in the southern part, which first in the west, now in the east can satisfy everyone; so that there is no place either toward the north, or toward the center of the day; neither toward the east, nor toward the west, that is not surprised to see what may happen. Moreover, I consider that the star of 1572 was close to the vernal equinox, which place according to the most learned astrologers signifies the state of religion, and this one appears close to the winter solstice, which place according to the aforesaid astrologers signifies the state of empires and kingdoms. Applying these things to our purpose I would like to believe that they may be premonitions of some great change, both in matters of faith and of kingdoms and empires, and since the first one was a little more similar to Jupiter, I believe that they must be premonitions of some happy state of affairs in the Christian and Catholic faith; Moreover, as the new star of the Savior Jesus Christ frightened Herod the tyrant and all Judea while at the same time He preached the conversion of the Gentiles, so it seems probable that these two new stars will disturb all tyrants and persecutors of the Christian and Catholic faith, and at the same time foretell the conversion of those who are now outside the assembly of the Holy Roman Church. In short, I believe that the two new stars are the precepts of that most happy state of affairs in the world in which Christ says in the Gospel that there will be one Shepherd and one fold. But since both have a Martian component, it is probable that this will not happen except with some great perturbation. Add to this the fact that this star of ours appears with the conjunction of Jupiter and Mars in which the great conjunction took place, which alone, according to the opinion of all the astrologers, may be the prelude of great changes in the Universe.

This is what I decided to say about this new star; I think it should be accepted by everyone, recognizing the loyalty with which I wanted to communicate what I have been able to observe so far about this wonderful event.

THE END.

Only a couple of years later Galilei gave a public reply, following a new attack by Capra, who this time had accused him of stealing the invention of the geometric compass. The proportional (or geometric) compass was an instrument capable of performing calculations up to third degree equations, useful for military and surveying applications, and its invention was attributed to Galilei. Capra claimed to have invented it, thus accusing Galilei of plagiarism.

Galilei then decided to defend his good name (and the economic income guaranteed to him by the compass, which was comparable to his salary), and wrote the pamphlet *Difesa contro alle calunnie et imposture di Baldessar Capra (Defense against the slander and imposture of Baldassarre Capra)*, in which he also spoke of Capra's three accusations related to the new star.

1. Regarding the alleged inaccuracy on the real date of apparition of the star, Galilei replied that he had said "approximately" during his conferences as a matter of prudence, since others could have seen the star in the days immediately preceding.
2. He pointed out that during the first of his three public lectures he had explicitly declared with words of praise that it was Capra who had been the first to observe the star in Padova. Galilei also dwelt on the irrelevance of the concept of primacy when it comes to discoveries, as he believed that being present during an event was just a fortuitous fact, that had nothing to do with the observer's skills.
3. Finally, with regard to the accusation of having observed a completely wrong area of the sky, Galilei replied that he had never spoken of "tail of the Swan" but rather of "tail of the Bear", thus accusing Capra not only of lack of attention during the lectures but also of poor knowledge of the Latin language.

Here is a translation of the part of the *Difesa contro alle calunnie et imposture di Baldessar Capra* by Galilei concerning the supernova of 1604.

From the *Difesa contro alle calunnie et imposture di Baldessar Capra* by Galilei

I have so far ignored the slanders and impostures written by Capra against me in the Astronomical Consideration about the New Star of 1604, published by him more than two years ago. Such a generosity from my side promoted his confident petulance. But since his misfortune has overcome my suffering, in order to make evident his oblique affection towards me, which has lasted for a long time, I will recount, also to relieve me of his other calumnies, what I have kept silent until now.

That bad affection of Capra's towards me began to germinate and to show itself with the appearance of the new star in 1604: before it had only spread its roots, nourished even by the putrid manure of his bad cultivator and consultant, or bad cultivators and consultants. He and his teacher, in order to practice, made every night astronomical observations; so in Padua they were the first to notice that new apparition, and they spoke about it to the Illustrious Mr. Iacopo Alvise Cornaro, Venetian gentleman, who passed me the notice. Capra was convinced, as far as I believe, that without their news I would not have succeeded in making the three long lectures that were attended by more than a thousand listeners. He and his master believed that what good I said was due to their advice, and that I for myself, without their lectures, would not have been fit to speak on such high matters. However, their warnings were nothing more than to learn from third persons that they had been the first observers of the appearance of the new star; if this record is to be held in such high esteem, it would be good that those who in the mathematical sciences aspire to some noble degree of glory spend all the nights of their lives observing with great vigilance above the roofs whether any new star appears, waiting for a favorable event. I knew very well that this was one of the highest praises that Capra had earned in the whole course of his mathematical studies, and therefore I did not want to defraud him of that merit: so in my first lecture on the new star, in the presence of him and his master, I said with words of praise that it was they who had been the first observers in this city. Therefore I was quite surprised by the complaints against me in his book on the new star.

But please note how much the desire to point out my actions, even if unreasonably, advances in Capra the will to lay charges that stain my honor. Since he cannot deny that I did attribute to him and his master the pride of having been the first in Padua to observe the new star, he does not talk about the honest mention I made of it, and imposes on me the lack of naming the Illustrious Cornaro, who was but a simple ambassador of what Capra had told him he had observed together with his master. See what Capra writes on this subject in his book on the new star, on p. 7b, where he concludes with these words: "From this we draw a necessary conclusion, namely that the excellent Galileo had the time and the place to observe this new wonder thanks to Cornaro, whom he did not mention in his lectures". But if I have named Capra and his master, of whom I was made aware through Cornaro, why reproach me for not having named the said gentleman?

But in order to follow my intention, which is to show by what impostures, partly frivolous and partly false, Capra has endeavored ever since to degrade my honor and reputation, consider first the uncivilized and inconsiderate manner in which he operates: to make himself capable of tearing me apart, he dares to print what he imagines I have said in my lectures and what I did not publish in print. It is therefore necessary to be very careful in speaking in the presence of these persons, who, as if they were spies of the world, pick up very subtly what others, either carried away by the course of words, or through inadvertence, or even through ignorance, let out of their mouths, and make it reach the ears of all. Will the privileges and abilities that time has granted to scholars, to be able to notice errors, correct them, once, twice and a hundred times revise, polish and punish their writings, be abolished and annulled by the petulant and vigilant censorship of these people? I do not know in which schools Capra learned this ugly custom: I do not think from his German teacher, because, being Tycho's scholar, he could learn from him, and show to his disciple what terms to use in publishing not only things said by others, but things communicated and sent around in private writings. Both could have learned modesty from Tycho, who, wishing to include in his writings some remarks of a still living friend of his about the new star of Cassiopeia, first asked his consent, and then added these words as an excuse: "I mix some things, even if not previously published, because the author himself has gladly granted it to me in letters"; and he did not use these words to blame or contradict him.

But why should I doubt that Capra knew that these actions of his were of the worst kind? On the contrary, it is also clear that he considered it an act of malice to discuss things already printed and published by others, complaining at the beginning of his astronomical internships of the fearfulness of critics, and writing these words: "Since in this tragicomedy of life is a catastrophe of human misery, that if someone, out of a desire to help mortals, or driven by his friends, did something to publish the law," etc.. But what is a sight of a mole in his own faults, is of an eagle and of a serpent towards the operations of others, if our mind is clouded by affections and interests! This poor man blames the corruption of this century of ours on the vigilant snares of the critics, who always in the guise of rapacious vultures are ready to pounce on the new parts that have just emerged from under the feathers of their fathers, and to maul them with their biting rostrums, and to beat them with their stinging claws, so that, oppressed by them in their first flight, they cannot open their wings to the sky and enjoy the wide fields of popular aura! And he does not notice how he, spurred on by even more ravenous lusts, pierces the nests of others, and breaking the bark of the unborn, tears the young, whose tender limbs, to be better formed, strengthened and consolidated, were still brooding under the beloved warmth of the patient father. Capra therefore blames others for his bitterness against the works already printed by their own authors, and tolerates in himself the impatience of not being able to wait for me to print mine; indeed, driven by the eagerness to tear them to pieces, impatient and afraid of losing such beautiful opportunities, he boldly resolves to publish them and then break them into pieces.

Judge for yourselves readers: this is a very great audacity, but it is also a small one, tolerable and excusable, if compared to the immense and not at all exemplary temerity used against me by this man, who, having heard nothing to justify his bitterness in my lectures, and yet wanting to tear me to pieces, wrote that I had said things that never came out of my mouth, as I will show you later. And mind you that I will not give you as the main argument for his malignity what he says on p. 5 of his *Astronomical Consideration,* attributing to me with great prominence and very unjustifiably, and only as far as his malignity is concerned, that is the facts that I have not stated about the time of the appearance of the new star, and that I have mistakenly said that it was located at about 18° of Sagittarius with almost two degrees of boreal latitude. On p. 6 he then attributes to me as a serious defect the confusion of the eighth day with the ninth and tenth, so that I did not know whether the star appeared on the 8th, 9th or 10th; adding that this should have been described with diligence, and again taking up the fact that I had not precisely determined its position with respect to the ecliptic. These inaccuracies, even if they had been true, were very slight and not necessary to the purpose of my lectures, which was only to show that the new star was outside the sphere of the elements.

These accusations would demonstrate much more the misery of Capra than the doctrine of my lectures; but being in more false, besides that immodest, denote its falsity and its recklessness. I have inaccurately stated the day of the first appearance of the star, and in fact the first words of my first lecture were these: "On the tenth of October of this 1604, a new light was seen for the first time in the sky."[83] It is true that little later, having spoken of the conjunction of Jupiter and Mars, which was on the 8th, and having to say that on the 10th the star was seen, I said: "It was also observed on the 8th, actually on the 10th"[84] correcting immediately the slip. And these were the confusions about the day of its first appearance: a lack that for its smallness shows how great is the malice of those who noticed it. So far for the location, I do not know why, in a discursive reasoning, in which according to me there was no need to cloud the mind of the listeners with degrees and their fractions, I thought it was better to say: in about 18° of Sagittarius, with about 2° of latitude, instead of saying: in 17° and 41 min of Sagittarius, with 1 degree and 51 min of boreal latitude. But if you have to be so severe in criticizing these clarifications, why Capra did not start to resume Tycho Brahe and many other famous authors, whose writings are recorded by him in the *Progymnasmata,*[85] who are so unprejudiced in assigning the place and time of the appearance of the star of Cassiopeia? Then the Illustrious Prince Wilhelm Landgrave of Hesse, as we read in Tycho Brahe's *Proginnasmata,* p. 491, sending to Tycho his observations on the new star of Cassiopeia, writes: "In the year 1572, on December 3, on the advice of the Elector of Saxony, I saw and observed for the first time a new star, bigger and more luminous of the same Venus, in the constellation of Cassiopeia."[86] And in investigating the position of the said star, one notices in reporting its right ascension, and in establishing its declination by means of the many observations made by the same Prince with precise instruments, differences in the ascensions of more than two degrees, and in the declinations of about 37 min:

---

[83] In Latin in the text.

[84] In Latin in the text.

[85] *Astronomiae Instauratae Progymnasmata,* a posthumous work (1602) by Tycho Brahe edited by Kepler, which includes the *De nova stella anni 1572*. The word Progymnasmata indicates manuals used in ancient schools for rhetorical instruction, full of exercises.

[86] Tadeás Hájek, Bohemian mathematician and astronomer graduated in Bologna, personal physician of Emperor Rudolf II.

- Thaddeus Hagecius, Bohemian, in his book entitled *Dialexis de novae et prius incognitae stellae...*, in assigning the time, says he first saw it around the Nativity of Our Lord.

- Gasparo Peucero,[87] in a letter of his dated December 7, 1572, writes: "I present the fact that a new star was born, which we saw in the fourth week under the constellation of Cassiopeia..."

- Paulo Hainzelio[88] writes[89]: "I saw this light for the first time on November 7 in the tenth house."[90]

- Michael Maestln writes: "In the year before 1572, the first week of November, a new star began to appear in the constellation of Cassiopeia, reaching the end of the Galaxy."

- Cornelius Gemma[91] writes[92]: "This star began shining on November 9."

- Girolamo Munosio, spanish mathematician professor in Valencia, does not write about the time more precisely, except that he knew[93] "with certainty that on November 2, 1572, this star had not yet appeared."

- Brahe himself does not make sure to state anything else, except that it began to be seen[94] "towards the end of the year 1572, in the month of November, towards the beginning, or at least in its first triad."

And on the position of the same star we find, in the same authors, differences of many minutes.

But since the place of the new star was not yet known to me with such precision when I gave my lectures (not many days had passed since its appearance), is that a reason for reproach that I did not want to determine its position up to the minute of arc? Or should I rather be praised for not having dared to give to the measurement the precision that cannot be obtained without a diligent observation repeated many times? This is evident from reading the differences in the positions assigned to the new star in 1572 and to this one by different observers. But, Immortal God, how can Capra accuse me of negligence for the lack of precision on a star that appeared the day before yesterday, while does not condemn his great ignorance in writing the apparent diameter of the Moon, measured by thousands and thousands? Because he, on p. 9, says that in the sky it does not occupy more than half a degree, i.e. 30 min; yet it is known from the books of all astronomers how the Moon in some days of the month occupies 30, 31, 32, 33 and 34 min, and sometimes even less than 29. This is really an unforgivable mistake, and it demonstrates great ignorance. Nor less wrong will be what he writes on p. 20, saying: "But above all the reasons for not having this star any parallax, it is very obvious that it cannot be except among the fixed stars, in which place the parallax is not perceptible because of its smallness." We therefore attribute to the fixed stars a parallax; he does not realize, nor understands, how in the fixed stars there is neither parallax nor can there be parallax, being those the last and highest visible bodies, with respect to which the inferior planets, and very near to us, show the diversity of appearance, called by astronomers parallax.

And these things, discreet readers, are not the main argument to convince you of the minimal science and of the supreme arrogance exhibited by Capra in his book on the new star. I invite

---

[87] Caspar Peucer, Slavic physician and mathematician.

[88] Paul Hainzel, German astronomer and regent of Augsburg.

[89] In Latin in the text.

[90] In astrology the sky is divided into twelve houses identifiable as segments within the circle of the zodiac.

[91] Astronomer in Leiden.

[92] In Latin in the text.

[93] In Latin in the text.

[94] In Latin in the text.

you to listen carefully to what I have to tell you about what he writes in the same book: although what he tells has nothing to do with his purpose, but is introduced only to hit me, he writes that I said in my lectures that the new star was in a straight line with the light of the Corona Borealis and with the light in the tail of the Swan, and then goes on to condemn as imperfect and useless the way I said I had ascertained the immobility of the said star, because it always maintained the same straight line with two fixed stars. Now, I never said that the new star was in a straight line with the brightness of the Crown and with the tail of the Swan, but with the brightness of the Crown and with the first of the three stars in the tail of Helix; but he believed that Helix means Swan and not Ursa, and what was an error of his ignorance, he wanted to attribute it to my fault and inadvertence. And that I never put the new star in a straight line with the Crown and the Swan, besides the testimonies I could produce of many who were present at my lectures, and who still have memory of them, is demonstrated by the copy in my possession of a kind of summary of my lectures, written in the form of a letter from the Reverend Antonio Alberti, Archpriest of Abano, to Giovanni Malipiero on 17 December, that is two months before the publication of Capra's book. Of this letter I will then transcribe the part that interests the present purpose, verified and authenticated as you will see at the end of this discourse.

But what really makes Capra's shamelessness evident is what follows.

A month before printing his book, Capra came to the illustrious Iacopo Alvigi Cornaro, and left him two questions written on a piece of paper, so that he could ask them on his behalf. Immediately Cornaro came to see me together with the excellent Francesco Contarini,[95] a gentleman of the noblest manners, and in addition to his knowledge of law, philosophy and sacred theology, a most gracious writer of Tuscan poetry, and he brought me the paper with the questions, which I still keep. The precise words used in the questions are these: "I doubt whether it is right to say that the new star always forms a straight line with the light of the Corona borealis and with the light of the tail of the Swan; and if the said stars (or even others that were) form a straight line, how is it possible that the straight line is preserved when the new star varies its height?"

I replied to these gentlemen that I was not surprised that Capra found strange this way of observing the immobility of a new star by referring it to two fixed stars with which it is in a straight line, since he was still young and a beginner in these studies; but I said I was surprised that the technique was not known to his teacher, without whose knowledge it was not credible that Capra had made the investigations, since of a similar way of observing there are just under 50 examples given by Ptolemy in Chap. 1 of book 7 of his *Almagest*. I added that I could excuse the said master for not having understood this in Ptolemy, whose lessons, being very difficult, are not for everyone; but I still could not excuse them for not having studied a similar way of observing in Tycho Brahe, whose techniques have become doctrine, and much celebrated in Michael Maestlin's writing of 1572 on the new star. He observed its position, and demonstrated its immobility and lack of parallax, only with a wire, always finding it in a straight line with two pairs of fixed stars. I suggested to tell Capra to study p. 54 of the *Progymnasmata* by Tycho Brahe. As for the other part of the question, I answered that it was false that the new star was in a straight line with the Swan and the Crown, but I told them that it was in a straight line with the Crown and with the first of the three stars in the tail of the Big Dipper, also called Elice; and I showed on my celestial globe how the same maximum circle passed through the position of the new star and through the Crown and the tail of Elice, adding that the same was in the same maximum circle as in the same straight line.

This answer of mine was reported by Cornaro to Capra, but without any profit because of his imprudence and ignorance; so that he was not dissuaded from printing, a month later, the book already prepared, with the same charges against me, persisting also in asserting that

---

[95] Natural son of the noble Taddeo Contarini, he was a teacher in Padua.

I said that the new star was in a straight line with the Crown and the Swan; and persisting in the same obstinacy, that to observe the site and the immobility of a star by referring it to others with which is in a straight line, is, in spite of Ptolemy, and before him of Hypparchus, Aristillus and Democare, and after him of Tycho, Maestlin and countless others, a fallacious and imperfect way. What unheard arrogance, what willful ignorance! Now what defense will we have against his slander every time he wants to impose some lie on us, since not only the fact that I have not said a folly, but the fact that we answer him with the intervention of more than one witness that I have neither said nor imagined it, is not enough to curb his foul-mouthed pen that makes false and arrogant accusations? Readers should note the carelessness of this man combined with malice, since he imagined that he could make others believe that I too have improperly misunderstood the Dipper, also known by oxen, or at least by cattlemen,[96] and said that I have mistaken it for the Swan, a constellation not less distant and different from how different a bear and a real swan are. But let's clear up the location issue, and then we'll move on from there.

Excerpt from the letter
of the Reverend Antonio Alberti, Archpriest of Abano,
written on December 17, 1604.

But it is also clear for the following reasons, that it cannot be within the seventh sphere. First, if it were in the region of the elements, as if it were in the highest part, it would have a different aspect; which is not, because the excellent Mr. Galilei has diligently observed it in a straight line with the first star of the three in the tail of the Big Dipper and with the bright star of the Crown, and he always found it in that straight line, etc.

Padua, April 15, 1607.

I, Giacomo Alvise Cornaro, affirm and testify that about a month before Baldassarre Capra printed his treatise on the new star, he gave me two questions on a piece of paper, so that I could show them on his behalf to the mathematician Galileo and get an answer. These questions were: first, whether it was well said that the new star formed a straight line with the tail of the Swan and with the light of the Northern Crown; and second, how sure was this way of knowing the site or the motion of a star by observing with which others it was in a straight line, since it was not possible to maintain the same straight line by varying the height of the new star. To which the said mathematician replied that as for observing the motion or site of a star, that is where it is, and if it has a motion away from the fixed stars, to see with which fixed stars it forms a straight line was a very safe way used by Ptolemy and other astrologers before and after him; and moreover he showed me and gave in note the passage of Tycho Brahe, in which he declares excellent the rule of Maestlin, who with a thread observed and found the site of the new star of 1572 putting it in line with other stars. And about the other question he answered that the new star of 1604 was not in line with the tail of the Swan and the Crown, but with the tail of the Bear and the Crown; he showed me that this was true on a celestial globe; and Mr. Francesco Contarmi also heard all this, and all this was duly reported by me to the said Capra the following day. In witness I testify by my own hand, sealed with my seal.

---

[96] As noted with regard to the Dialogo di Cecco di Ronchitti, in the Paduan dialect the word has a derogatory sense.

I, Giacomo Alvise Cornaro, confirm what written above.
I, Francesco Contarini, son of Taddeo, was present,
and affirm that what is narrated above is true.

Capra condemns in his booklet the way to investigate the immobility of a star by observing
if it always remains in a straight line with two identical fixed points, and says: "This way is
not safe, because, even if when the new star was quite high, it formed a straight line with two
supposed fixed points, then near the horizon for the refraction of vapors it could not make a
straight line".

Anyone who has a mediocre understanding of the first principles of astronomy will clearly
see how Capra does not understand this way of observing the immobility of a star, which
he takes as fallacious. Capra believed, as his words show, that I and the other astronomers,
having observed three stars in a straight line to ascertain whether any of them has a motion
of its own, return a few hours later to observe again whether they maintain the same line,
in which finding, since a fallacy may occur with respect to refractions and to the fact that
the said stars have changed their position above the horizon, a certain science should not
be established. But who told you, Mr. Goat,[97] that between one and another observation
several hours must pass? Who can be so clever as to believe that even the motion of Jupiter,
not that of Saturn, or of another celestial body, if found, slower, can be detected by such
short observations? It takes not hours, but days, weeks, months, years and even centuries
between one observation and another, before it can be established with certainty that a star
does not have a different motion from the others. Ptolemy stated that fixed stars do not change
their mutual position: on what basis? Because all the mutual positions were measured many
hundreds of years before. by Aristillus and Timocare, and then by Hipparchus, along the
same alignments; and I said that the new star did not show to have a proper motion, because,
having found it from the beginning in a straight line with the two said fixed stars, many
days and weeks later, and not a few hours, it showed the same alignment. What do these
observations have to do with refractions? And who forbids me to make observations when
the star is at the same height above the horizon? So take it up with your lack of knowledge
and comprehension, and not with the observations excellently made by me and before that
by other astronomers.

I believe, my judicious readers, that I have demonstrated very clearly the malevolent disposition of Capra towards me. This malevolent disposition began to be evident many years ago,
and without any restraint of modesty it overflowed with audacity into the false impostures
against me, which you have so far taken up. Now here I leave you to think what slanders,
curses and snares may have been spewed and stained against my reputation, openly and
secretly, by him and his miserable advisers, living 14 or 15 years in the same city, and seeing me every day. If with such falsehood and impudence he did not hesitate to publish the
aforementioned impostures, in such a way that he could be sure that they had reached my
ears, what do you think were his accusations in his private discussions, and what slanders
did he think he could secretly make by reporting them verbally?

It may seem impossible to some that in Capra's soul an intestinal hate of me should have taken
root so firmly, without me having given him some serious occasion to do so, by offending
him, his father, or some of his relatives, either by deed or by word, and that my natural
enmity of ignorance alone should have been sufficient to provoke his so bitter anger. I do
not want to remember that I have spoken to them no more than three or four times in all the
years that I have been in this city, and that only for their benefit. And I believe, if I remember
correctly, that the first meeting was to entrust his father with the task of teaching fencing to
the illustrious Count Alfonso di Porcia, a gentleman from Friuli. The second time was with

---

[97] "Capra" means in Italian "Goat", a term also referring to ignorant people; here Galileo plays with
the word.

his father and him in the home of the Illustrious Cornaro, begged by them to show him my compass and some of its operations, as is seen below in the certificate of the same Cornaro. The third time, hearing that there was in the hands of the illustrious Orazio dei Marchesi del Monte an order from a very great Prince to know a certain secret, and that no expense should be spared, and that the said gentleman came to inquire from me whether I knew a man named by the said Prince as possessor of the desired secret, I told him I did, but that he was not then in these parts. So I approached Aurelio Capra, father of Baldassarre, to ask him if he knew the said secret, and if, having it and being able to receive from a very great Prince a very great recognition, he wished to communicate it to him. He answered that he did; and I went at once to see Signor Orazio, telling him that I had found another who possessed the desired secret, and that I thought the Prince did not care much to know the secret more from that person named by him than from others. Horatio thought the same, so I took him to Capra, and I believe this affair was accomplished to the satisfaction of both parties.

This is what I remember having to do with these people; judge for yourself whether I deserve to be treated so badly by them. But why should I want to justify myself with other depositions that I have never offended him? What fuller testimony must I seek in confirmation of my good affection for him than the tolerance he had from me for more than two continuous years, that his Astronomical Consideration, in which he persecuted me so falsely and maliciously, remained unanswered, since I could so easily purge myself and show the world his falsities? I never wanted to do this, and I would never have done it, if his obstinate, incomparable and incomprehensible fear had not finally won, or rather forced, my suffering by this last action of his. And not only have I wished to refrain from answering and discovering his trifles and malignities, but (and perhaps with greater note to my reputation, than with praise to my indulgence) I have forbidden the printing of a letter, written by a pupil of mine in my defence concerning the calumnies and ineptitudes of which Capra has advanced against me in the said *Astronomical Consideration*. This polite apology was composed immediately after the publication of the said *Consideration,* and in bringing it back to me from the said pupil to see it again, I kept it close to me, and I still have it, and I did not wish it to be published out of compassion for the young Capra, and also hoping that his father or other of his friends, without so much blushing, would correct and for the moment modify his arrogance. And so that no one will believe that what I have now said is a fiction, the above apologetic letter presented by me before the Illustrious Lords Podestà and Capitanio[98] of Padua will be cited at the end of this defense, and will be seen, acknowledged and authenticated by their Illustrious Lords; all the other writings and attestations made in Padua, which I will produce in this defence, and of which the originals will remain in the chancellery of the Illustrious Podestà, to be shown to those who wish to see them, will also be named and authenticated; and the other attestations that I will produce and that are made in Venice will also be named and authenticated, presented in their originals, and acknowledged by the Illustrious Lords Reformers. The originals of these will remain in the chancery of the Illustrious and Exalted Lord Reformer to be shown to those who wish to see them; and the other testimonies that I will produce and that are made in Venice, once the originals have been presented and acknowledged by the Illustrious and Exalted Lord Reformers, will be similarly authenticated by their Lordships.

This human and long suffering of mine, this concealment of the rude affronts made to me by Capra, which in anyone else would have finally, by the remorse of conscience, aroused repentance for his faults, and softened any bitterness which, being rooted in his taste, made him feel nauseous when he reflected on any of my actions, has on the contrary so inflated his vain folly, promoted his arrogance, and enlivened his boldness, blunted his modesty and sharpened the poison of all his senses, and especially of his tongue, but above all (and this by God's grant) so dimmed every light of his mind, and took from him, by his punishment, all judgment and speech, that, regarding my forbearance as timidity, my dissimulation as

---

[98] Regent of Padua on behalf of the Republic of Venice.

stolidity, my contempt for his foolishness as my crass ignorance, has allowed himself to be drawn into this latest abominable, infamous and detestable operation of his, in which he has believed and persuaded himself that he is able not only to defame me, but to mock and deceive many other people who are well aware of the truth of the matter. And here you will forgive me, pious readers and lovers of right, if perhaps with too much tedium I shall keep you occupied in reading this defense of mine; and you will excuse me if I still take too much time in evidencing the errors of this man, so that his ignorance may not cost him recklessness and folly.

[...]

He was able to prove that Capra had copied his work regarding the compass, while he admitted that Capra had the priority regarding the observation of the supernova. Capra was condemned by the court of Venice.

Other Treatises on the Supernova in Italy

Van Heeck and the De Stella Nova Disputatio

The Accademia dei Lincei, perhaps the first academy dedicated to the experimental study of nature, thought it appropriate to express an opinion on the new star, and the Dutch physician and astrologer Johannes van Heeck was charged with the task of writing a report. Together with Federico Cesi, Anastasio de Filiis and Francesco Stelluti, van Heeck (1574–1620 approx.), who spent most of his maturity in Rome, had been one of the four founding members of the Academy.

Van Heeck reaffirmed the orthodox Catholic positions. He concluded that the supernova of 1604 did not show any sign of parallax, which meant that the event could not have taken place near the Earth, but had to be located much farther away, between the fixed stars of the firmament or perhaps even further away—on this he agreed with Galileo, Kepler and other astronomers. However, he reconciled this conclusion with the Aristotelian model by arguing that the apparently new star was not new at all; although it had been there since creation, like other stars, it was only occasionally visible, when a rare and transparent part of the sky, almost a hole, passed by it. He accused those who claimed that the sky was mutable of being in contrast with the Holy Scriptures, and attacked Tycho Brahe for his religious beliefs and alleged scientific errors. Expanding his polemic against all erroneous ideas, he attacked Protestantism. The language and tone of van Heeck's text were very intemperate. He rebuked the babbling new philosophers for their profane ignorance and their stupid ostentation in departing from Aristotle.

The pamphlet, sent to Cesi in January 1605, posed a number of problems. Cesi held Kepler in high esteem, and therefore modified van Heeck's text, removing everything that was hostile to him or to other astronomers and to Protestantism. He also removed most of the defense of Aristotelian cosmology, because it was important that the Academy aligned itself with the new astronomical discoveries and not entrenched in the defense of Aristotle. He then published the pamphlet without the emblem of the lynx of the Academy on the title page.

The pamphlet, not particularly significant (for better or worse, since Cesi had blunted it abundantly), was published in Rome by Zanetti in February 1605, and received little attention. Van Heeck was upset for these editorial changes, which had been undertaken without his knowledge or permission.

The Discorso di Lodovico Delle Colombe

In 1606 was published the *Discorso di Lodovico Delle Colombe nel quale si dimostra che la nuova stella apparita l'ottobre passato 1604 nel Sagittario non è cometa, né stella generata, o creata di nuovo, né apparente: ma una di quelle che furono da principio nel cielo; e ciò esser conforme alla vera Filosofia, Teologia, e Astronomiche dimostrazioni* (*Discourse of Lodovico Delle Colombe in which it is shown that the new star appeared last October 1604 in Sagittarius is not a comet, nor generated star, or created again, nor apparent: but one of those that were from the beginning in the sky, and this being in accordance with the true Philosophy, Theology, and Astronomical demonstrations*) in which he defended an Aristotelian view of cosmology after Galileo Galilei had used the opportunity of the supernova to challenge the Aristotelian system.

Delle Colombe assumed the incorruptibility of the heavens in spite of the new apparition. He assumed that the star had always existed, but it was small, and because of the huge room it became visible only when a part of the crystalline sky, which was denser than the rest, passed in front of it and enlarged it like a convex eyeglass (an opposite explanation to that of van Heeck).

The book also contained an attack to Galilei on a philosophical rather than physical basis. The line of attack is as follows. Galilei explained the meaning and relevance of parallax, reported that the new star had none, and concluded with certainty that it was beyond the Moon. Here he could have stopped, having shot his only arrow. Instead he sketched a theory that ruins the Aristotelian cosmos, saying that the new star consists most likely in a large amount of airborne material that came out from the Earth and shone by reflected sunlight, like the Aristotelian comets. Unlike them, however, it could be located beyond the Moon. Not only did it bring change to the heavens, but it did so provocatively by importing corruptible terrestrial elements into the pure fifth essence. This would lead to upheaval in the heavens. Interstellar space could be filled with something like our own atmosphere, as in the physics of the Stoics, which Tycho had referred to in his lengthy account of the supernova of 1572. And if the material of the firmament resembled that of bodies down here, a theory of motion built on experience with objects within our reach could apply to the celestial regions as well.

The attack aims to hit a very important aspect for Galilei: the possibility to formulate the discussion of physical phenomena in mathematical terms. But mathematics was a discipline of which the members of what Galilei called "the league of the pigeon" (a nickname coined by Galilei himself for Lodovico Delle Colombe and his friends: piccione, "pigeon", is in Italian synonimous of "colomba", but also refers to naive people, or to people caring about useless things) showed a total ignorance.

A few months after the appearance of Delle Colombe's book, a reply was published in Florence under the title *Considerazioni d'Alimberto Mauri sopra alcuni luoghi del discorso di Lodovico Delle Colombe intorno alla stella apparsa nel 1604*. Alimberto Mauri was a pseudonym (Mauri is otherwise unknown) and Delle Colombe (like many scholars since then) believed that the author was Galilei, although according to critics the attribution is much more dubious than that of the 1605 dialectal pamphlet. The book ridiculed many of Delle Colombe's opinions about the star, and belittled him as "our pigeon." It asserted that astronomy did not need Aristotelian philosophy, and should focus on observation and mathematics. The approach of first seeing something in the sky and then developing an elaborate explanation to match that observation with Aristotelian cosmology was stupid.

Delle Colombe answered Mauri by publishing in 1608 a booklet entitled *Risposte piacevoli e curiose di Lodovico Delle Colombe alle considerazioni di certa maschera saccente nominata Alimberto Mauri fatte sopra alcuni luoghi del discorso del medesimo Lodovico dintorno alla stella apparsa l'anno 1604 (Pleasant and curious replies by Lodovico Delle Colombe to the remarks of a certain know-it-all mask named Alimberto Mauri made over some places in the speech of the same Lodovico concerning the star that appeared in the year 1604)*. In this text he not only attacked again the ideas of Galilei and those of Mauri, but he associated the ideas of Copernicus with those of Galilei, which was a very dangerous accusation at the time of the Inquisition.

Less Relevant Treatises

Raffaello Gualterotti, in his booklet *Sopra l'apparizione d'una nuova stella,* published in Florence by Giunti in 1605, claimed that the sky is penetrable, and that vapors coming out of the Earth had risen up to the eighth sphere to condense there in the form of a star (an idea that, as we have seen, even Galileo had caressed in 1604, and then rejected).

In Padua appeared in 1605, for the types of the publisher Pasquati (the same as the *Consideratione astronomica* by Baldassarre Capra), a small treatise entitled *Discorso sopra la stella nuova comparsa l'ottobre prossimo passato (Discourse over the new star that appeared last October),* by the "excellent astrologer and doctor Astolfo Arnerio Marchiano". Some identify Marchiano, otherwise unknown, with Galileo, but the hypothesis seems to me very remote: unlike the two peasants of Brugine, Marchiano does not seem to shine for astronomical culture, and launches into astrological predictions, which is not in the style of Galileo. The treatise underlines however the enormous cultural impact of the astronomical event: "between learned, and valiant professors of the worldly sciences implacable tumult, inextricable wander, and among them this Star has caused already discords, but perhaps seeds, and origins of deep and serious disdain".

Many other notes are only manuscripts. Among them there is also one (scientifically irrelevant) by Francesco Ingoli (1578–1649), already a student of Galileo and his future enemy in the inquisitorial process, entitled De stella anni 1604. Ingoli supported Aristotelian arguments similar to those of Lorenzini and Delle Colombe.

References

1. Alessandro De Angelis, *I diciotto anni migliori della mia vita*, Castelvecchi, Roma 2021
2. Stillman Drake, *Galileo Against the Philosophers*, Zeitlin & VerBrugge, Los Angeles 1976
3. *Dialogo de Cecco di Ronchitti da Bruzene in perpuosito de la stella nuova,* Merlo, Verona 1605
4. Marisa Milani (editor and translator), *Dialogo de Cecco di Ronchitti da Bruzene in perpuosito de la stella nuova,* Editoriale Programma, Padova 1992
5. Giampiero Bozzolato (editor and translator), *Dialogo di Cecco Ronchitti da Bruzene e Galileo Galilei a Padova,* Centro internazionale di storia dello spazio e del tempo, Brugine 1992
6. Lorenzo Tomasin, "Galileo e il pavano: un consuntivo", Lingua Nostra XXXIX, 2008, 22
7. Matteo Cosci, "Disputatio accademica e poesia", in *I generi dell'aristotelismo volgare nel Rinascimento,* M. Sgarbi ed., La filosofia e il suo passato 66, CLEUP, Padova 2018, and other works by Cosci
8. Antonio Daniele, *Intorno a Galileo,* CLEUP, Padova 2022
9. Cambridge Italian-English Dictionary, https://dictionary.cambridge.org

Chapter 4
Past, Present, Future

At the time of the appearance of the *stella nova*, Kepler was the imperial astronomer of the Holy Roman Emperor Rudolph II. In 1601, when he was only thirty years old, he had taken the place of his master Tycho Brahe which had designated him as his successor—the appointment had come to him two days after Tycho's funeral. He was Galileo's correspondent, and the letters between the two scientists express mutual trust and esteem. When Galileo left the Venetian republic, in 1610, he recommended Kepler as his successor to the chair of Padua—unfortunately the plan did not work.

Kepler studied the *stella nova* for more than two years, at the end of which published in 1606 the work *De stella nova in pede Serpentarii (The new star in the foot of Ophiuchus)*. This book is the most complete archive of information on the new star, and contains an enormous amount of observational data, acquired during the 18 months in which the star remained visible, provided by astronomers from all over Europe. In addition to the descriptive part, which includes very precise data on the position of the star with respect to other fixed stars, as well as its changes in color and brightness, there are also chapters in which Kepler could not avoid to address, because of his official role as imperial astronomer, topics very fashionable at the time, concerning the examination of possible meanings of the appearance of the new star, which was considered the bearer of omens. In particular, the astronomical event was associated with an omen of the spread of Christianity and an increasing power of the Holy Roman Emperor.

Galileo resumed his discussion on the new stars (both that of 1572 and that of 1604) in his *Dialogo sopra i massimi sistemi*, published in 1632, during the first and especially the third days of discussion, without hiding in this his Copernican thought nor the anti-Aristotelian one and the arguments on the distance of the new star, calculated through parallax, which was definitely larger than that of the Moon.

> Salviati— [...] we Italians are making ourselves look like ignorants and make foreigners laughs, especially those who have broken with our religion; I could show you some very famous ones who joke about our Academician and the many mathematicians in Italy for letting the follies of a certain Lorenzini appear in print and be maintained as his views without contradiction. [...]

A. De Angelis, *Galileo and the 1604 Supernova*,
SpringerBriefs in History of Science and Technology,
https://doi.org/10.1007/978-3-031-59486-1_4

You will be no less astonished at their [the Aristotelians'] manner of refuting the astronomers who declare the new stars to be above the orbits of the planets, and perhaps among the fixed stars themselves. [...]

In the case of the three superior planets, Mars, Jupiter and Saturn, it is argued that they are always very close to the Earth when they are at the opposition of the Sun, and very far away when they are at the conjunction; and this approach and departure is so important that Mars is seen 60 times larger when close than when at the conjunction with the Sun. Next, it is certain that Venus and Mercury revolve around the Sun, because they never move far away from it, and because of their being seen now beyond it and now on this side of it, as Venus's phases conclusively prove. [...].

But returning to the first general apprehensions, I reply that the center of the celestial conversions of the five planets, Saturn, Jupiter, Mars, Venus and Mercury, is the Sun; and it will be of the Earth's motion as well, if we put it in the sky. As for the Moon, it has a circular motion around the Earth, from which (as I have already said) it cannot be separated in any way; but this does not keep it to go around the Sun together with the Earth in its annual motion.

and writing at the end:

Salviati— Please, Sagredo, let's not worry any more with these particulars, especially since you know that our purpose is not to determine decisively or to accept as true this or that opinion, but only to propose for our own pleasure those reasons and answers which on either side can be adduced; and Simplicio answers this in redemption of his Peripatetics. Therefore we shall suspend judgment, and leave this in the hands of whoever knows more about it than we do.

He does not hesitate to draw a Copernican figure with the Sun at the center and the Earth revolving around it, a figure very similar to the one on page 41.

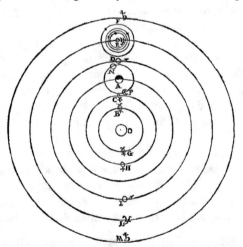

But let us go back to astrophysics. The analysis of the data collected by Kepler and Galileo, together with those of Chinese and Korean astronomers, allowed to identify the time profile of the emission in the first months of the *stella nova* as typical of a type Ia supernova. As we wrote in Chap. 1, inactive stars beyond a limiting mass, called the Chandrasekhar limit, which is about 1.4 solar masses, tend to collapse. If a "cold" white dwarf star (this is for example the destiny of our Sun) belongs to a binary system of stars (and this is not the case of the Sun), it could merge with its companion, exceed the Chandrasekhar limit and start to collapse, increasing its temperature up to the point of nuclear fusion. Within a few seconds from the beginning of the fusion, the conditions of very high pressure and very high temperature cause an uncontrolled thermonuclear reaction that releases enough energy to disintegrate the star in a violent explosion: in this case we have a supernova called of type Ia.

In 1941 a very important observation occurred. The astronomers of Mount Wilson Observatory, in California, discovered a very faint nebula (with a brightness of apparent magnitude 19, at the limit of visibility). It looked like a broken mass of nodes and filaments covering an area of about 4 min of arc in diameter (very large: to give an example, the diameter of the Moon seen from Earth is 30 min of arc, i.e., half a degree), reddish in color. In 1943 the German astronomer Walter Baade recognized it as the remnant of the supernova exploded in 1604, which was called Kepler supernova. The association made it possible to combine modern observations with the measurements of the astronomers in 1604 and 1605. It was discovered that the object is a strong source of radio waves and, later, also a source of X-rays, millions of times more energetic than visible light; recently also the emission of gamma rays, billions of times more energetic than visible light, has been revealed. This suggests that it is a site of acceleration of cosmic rays, charged particles that arrive on Earth from the cosmos. A recent image is shown in Fig. 4.1. At a distance of 16 000 light years from the Earth, the "supernova remnant" has a radius of about 10 light-years, that is 700 thousand times the distance between the Earth and the Sun. The expansion speed of the "bubble" is very large for a supernova of that age: about 10 000 km per second (one thirtieth of the speed of light).

As we have said, observations in the 1600s indicate that this was a type Ia supernova, the result of the collapse of a binary system of stars. We think that all such explosions are similar, with a few corrections that we think we know well: the mass of white dwarfs that explode as a result of accretion processes should always be the same, more or less, and therefore the amount of energy produced should be more or less the same. For this reason, type Ia supernovae are used as "standard candles" to measure their distance from us, as their apparent magnitude should depend almost exclusively on the distance at which they are located. But the cloud projected by the explosion of the 1604 supernova is expanding at a speed too high compared to what has been predicted and observed for other similar supernovae (as we said, in some places one thirtieth of the speed of light). This disturbs us a lot: most of our knowledge about the expansion of the Universe is based on the fact that Ia supernovae are standard sources of light. The supernova of 1604 could therefore reserve great surprises, as to our illustrious predecessors of four hundred years ago. We need new data to understand.

No one can say with certainty when and where the next great galactic supernova will appear; maybe it will be Betelgeuse, the brightest star in the Orion constellation. This red giant star not far from us will explode in an undefined time between now and one hundred thousand years in the future, and will become one thousand times brighter than Venus. Maybe it will be Aludra (Al-Ludra), the fifth brightest star in the

Fig. 4.1 The remnant of the
1604 supernova, in a *collage*
of images at various
wavelengths from large
NASA telescopes (in
particular the Hubble Space
Telescope) sensitive to X
rays, ultraviolet, visible and
infrared light

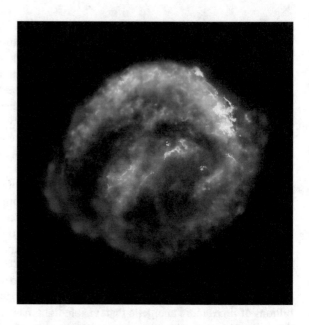

constellation of Cane Major (the constellation of Sirius). Whichever star becomes a
supernova next, we physicists would like the event to happen tonight.

At the end of his treatise, Kepler does not provide predictions about the date of
the next supernova, but goes so far as to make an astrological prediction for the
momentous historical event associated with the next great conjunction in a fiery
trigon, scheduled for about 2400.

Kepler wonders what will happen in 2400 to Germany so flourishing in the year
1600. "Who will be our successors? Will they remember us? All this, however, if the
word will still stand."

Postscript

This study started from a bibliographical research done by Selenia Broccio for her bachelor thesis, and I am very indebted to her; fundamental is the work of Antonio Favaro, curator of the National Edition of Galilei's *Opere*. It includes the exchanges of opinions involving Galilei in the months following the explosion of the supernova: his lecture notes (1604–1605) and some exchanges of letters, a summary and some extracts of a pamphlet written by Lorenzini in 1605 on behalf of the Aristotelians in response to the lectures of Galilei, and the Italian translation of the *Dialogo de Cecco di Ronchitti* (in Paduan dialect) written according to Favaro (and according to me) by Galileo with the collaboration of his student Spinelli as a response to Lorenzini.

I thank Cesare Barbieri, Andrea Bellin, Alessandro Bettini, Giovanni Busetto, Michele Camerota, Rosaria Candiloro, Graziano Chiaro, Marianna Giannicolo, Lorenzo Marafatto, Bruno Milani, Ivano Paccagnella, Carmela Patanè, Barbara Serino, William Shea, Andrea and Nadia Sitzia, David Speranzi, for their suggestions and their help in the bibliographic research.

My work has been partly carried out in Udine at Palazzo del Torso Antonini, today seat of CISM (Centre International des Sciences Mécaniques).

This book was written as part of the celebrations for the eighth centenary of the University of Padua, and is dedicated to this anniversary. *Universa universis patavina libertas*.

© The Author(s), under exclusive license to Springer Nature Switzerland AG 2024   77
A. De Angelis, *Galileo and the 1604 Supernova*,
SpringerBriefs in History of Science and Technology,
https://doi.org/10.1007/978-3-031-59486-1

Printed in the United States
by Baker & Taylor Publisher Services